■

U N T H I N K I N G Thinking

Purdue
University
Press

West
Lafayette,
Indiana

\U N T H I N K I N G Thinking/

Jorge Luis
Borges,
Mathematics,
and the
New Physics

Floyd Merrell/

Copyright © 1991 by Purdue Research Foundation, West Lafayette, Indiana 47907. All rights reserved. Unless permission is granted, this material shall not be copied, reproduced, or coded for reproduction by any electrical, mechanical, or chemical processes, or combination thereof, now known or later developed.

Permission to reprint the Escher prints on pages 73 and 110 from Cordon Art Bv. is gratefully acknowledged.

Library of Congress Cataloging-In-Publication Data

Merrell, Floyd, 1937–
    Unthinking thinking : Jorge Louis Borges, mathematics, and the new physics / by Floyd Merrell.
       p. cm.
    Includes bibliographical references and index.
    ISBN 1-55753-011-4 (alk. paper) :
    1. Borges, Jorge Luis, 1899– —Philosophy.
    2. Mathematics—Philosophy. 3. Physics—Philosophy.
    4. Science—Philosophy.
    I. Title
PQ7797.B635Z779   1991
868—dc20   90-20128

Book and jacket designed by J. Marc Peterson/studio PitchFORK

Printed in the United States of America

Contents

vii	**Key to Quotes from Borges's Works**
ix	**Preface**
xvii	**Acknowledgments**

Chapter One
1	**Intellection and Contemplation:** **An Impossible *Coniunctionis Oppositorum***
2	**1** Metaphysics of Deceit
16	**2** Where Fiction Ends

Chapter Two
31	**A Predilection for Paradox**
32	**1** According to the Eye of the Beholder
42	**2** To Reach the Unreachable

Chapter Three
53	**The Demise of Totalizing Quests**
54	**1** Models of Infinity
67	**2** The End of Certainty

83	**Interlude**
84	**1** Number Power/Word Power
91	**2** The World as Dream

Chapter Four
103	**The Universe as Library**
104	**1** The Fearful Sphere
117	**2** Parmenides Triumphs

Chapter Five
133	**Chronos in Chains**
134	**1** Time's Eternal Struggle
145	**2** Singularities and Other Strange Phenomena

Chapter Six
155	**What Is Real?**
156	**1** The Most Probable World
170	**2** A Throw of the Dice

177	**3**	Multiple "Realities"
182	**4**	Quantum—and Textual—Interconnectedness
198	**5**	Symmetries, Mirrors, Broken Symmetries

Chapter Seven

209 **Suspended within Language**
211	**1**	Language against Itself
228	**2**	The Rules of the Game
234	**3**	Texts of Our Own Making
241	**4**	The Receding Horizon

245 **Notes**

265 **Works Cited**

291 **Index**

Key to Quotes from Borges's Works

A	The Aleph
B	"La bibliteca total"
BD	Chronicles of Bustos Domecq
BI	The Book of Imaginary Beings
BS	The Book of Sand
CB	Cuentos breves y extraordinarios
D	Discusión
DT	Dreamtigers
F	Ficciones (English)
HE	Historia de la eternidad
I	Inquisiciones
IA	El idioma de los argentinos
L	Labyrinths
OI	Other Inquisitions
OP	Obra Poética
P	"Preface"
SN	Siete noches

I have quoted from English editions when available.

Preface

I begin this perhaps fallible preface with the parable about an imaginary scholar who, in the preface of her own book, concedes to the likely occurrence of occasional errors among her statements. Of course she has made a considerable number of assertions in her text, and, given each one of these individually, we must suppose that she believes them to be correct, but by her own admission, at least a few of them are in all probability false. The conjunction of her true assertions and the indeterminate number of false ones, then, makes up an inconsistent set, so in spite of our scholar's intellectual honesty, she has placed herself in a dilemma. Yet, since she can obviously bring herself to admit to such an inconsistent set of commitments and live with it quite comfortably, the "paradox" is not, we must surmise, unbearably pernicious.

Jorge Luis Borges, in an ironically similar vein, once declared that he is often asked what the message of a particular story is, to which he responds that it has no message. He continues with the disclaimer that he is "neither a thinker nor a moralist, but simply a man of letters who turns his own perplexities and that respected system of perplexities we call philosophy into the forms of literature" (P, ix). Of course, one's denial that one

has a message is itself a message, and the claim to be neither a thinker nor a moralist implies that one has thought, usually rather extensively, about moralism and the very act of thinking. If the contradictory character of Borges's statement does not diminish our respect for him, it is because we do not ordinarily perceive it to be malignant. We all occasionally make comparable avowals, generally ignoring their paradoxical nature and going on with the game, as the later Wittgenstein (1974, 303) would counsel us. In fact, it is advisable to do so, for the act of resolving a paradox can threaten to become itself a paradox. Borges, we must concede, has learned to live with his and a host of other Western World paradoxes. He makes wholesale use of them in his fictions, but, unlike most other spinners of paradoxes, he rarely abuses them. This is not surprising, since Borges tends to see in most philosophical endeavors not their truthfulness, their excellence of argument, or their ingenuity but their collective insufficiency, and he sees fit to disparage them accordingly (Butler 1973, 148).

In this light, "unthinking thinking" reveals the present inquiry's underlying strain. The title, I believe, is appropriately ambiguous. It implies either the project of unthinking traditional Western thought, or, paradoxically, thinking without there being any accompanying process of thought (an inevitably abortive attempt by sheer intellect to approach a mystical insight). Both interpretations are germane to the nature of Borges's work, as we shall observe.

Borges's most cerebral writings, chiefly found in the short stories collected in *Ficciones* (1944) and *El Aleph* (1949) and in his essays *Otras inquisiciones* (1952) and *Historia de la eternidad* (1936), might seem at the outset to be objective, rational, and detached, the result of a writer's forging for himself a personal *modus operandi* as he goes about his task of world-making. The work of the romantics, consequently, appears to be completely undone. Things that were close become distant, process is replaced by product, the particular becomes generalized. But on closer analysis, one discovers that this is not entirely so, for in Borges's case there inevitably comes a time when the intellect no longer finds answers to thought's harvest, and it turns inward to contemplate its own activity. Then, finally, at the crossroads where intellect and intuition meet, Borges's creativity commences, which is neither detached nor committed, neither objective nor subjective, but somehow both simultaneously. This cre-

ative drive squarely faces the infinite, and apparently chaotic, variety which is the universe, and it is condemned to oscillate between two poles: imagining a godlike possession, if only for a fleeting instant, of the totality, and at the same time realizing the futility of attempting actually to possess it (Barrenechea 1965, 77). Borges's fictions, indeed all fictions, can do no more than depict the inconsistency of the universe because any and all forms of knowledge can be no more than vague and approximate, doubts inextricably finding their way into the interstices. Yet Borges never fails to force a new understanding of the world.

How does he accomplish this? Perhaps, one might conjecture, in a manner similar to the Greeks' hypothetico-deductive method. To reason about the concepts of mathematics, they began with axioms, truths presumably so self-evident that they could not possibly be doubted, and from these axioms proofs were generated. This was indeed a radically novel mode of thought. If one were to measure the sum of the angles of two dozen triangles and find them to be 180 degrees in each case, one's resultant proof would be inductive, not deductive, and hence mathematically unacceptable. Deduction is much more stringent, more the product of unencumbered intellection, than induction. Nonetheless, Greek mathematicians and philosophers insisted on the exclusive use of deduction because it was thought to yield eternal truths. Borges, in contrast, creates by positing the equivalent of deductive hypostats while promising no truths; in fact, he customarily proceeds to demolish supposed truths, leaving the reader with little or no firm ground on which to stand.

Although André Maurois (1962, ix) remarks that Borges "has read everything that exists," the fabulist himself confesses that he gets more out of rereading than reading, and that he has read relatively little, for "nowadays there's so much written about books that one hardly finds one's way through to the books" (Burgin 1968, 101). The casual reader of Borges's work might also be prone to conclude that he must certainly have read everything; yet there are no more than occasional glimpses here and there of a few worn relics from the archeological museum of scripts. His works invariably point beyond, to some underlying order or substrate that serves as a conjectural cornerstone upon which his narrative is built. It is significant, in this light, that Borges's stories generally contain neither character nor plot in the traditional sense. Like science fiction, they are constructed

from root propositions, contrary-to-fact conditionals. The author seems to be telling us that if X were the case, then most likely Y would ensue. This is, properly speaking, analogous to the hypothetico-deductive method, in contrast to, say, a romantic poem, which in an empirical sense attends to particulars. Yet, once again, Borges is not merely a pure deductivist of sorts, for he delights in fusing dummy authors with existing ones, mixing fictional characters with real people, and juxtaposing imaginary books, places, and times with historical ones. In the face of these unexpected conjunctions, all elements seem to acquire a "reality" as "real" as the "real" itself, and Borges apparently expects, even encourages, the reader to build on his hypothetical constructs, for "[i]f people find more in a story than what I intended to put there, well, that's all to the good, because I think any story or any poem should have more in it than what was intended by the writer; because if not, it would be a very poor one" (Christ 1972, 399).

In spite of Borges's penchant for constructing hypostats, by his own admission he is not a systematic thinker.[1] He has read much philosophy, but his reading is eclectic. In fact, a large portion of his booklore is the reverse of that of the scholar: he enjoys encyclopedias. Borges's flights of creative insight derive from casual, diverse perusal. He delights in radically novel ideas not as a rigorous thinker and much less as a system builder but as an artist, for their aesthetic value. He resorts neither to religious belief (though he would like to believe) nor to pompous metaphysical claims (though there persists a longing for Truth), nor is he a nihilist (he is saved from it by his playfulness). Consequently it becomes at times practically impossible to disentangle the diverse metaphysical postures that Borges at any given moment may assume (Amaral 1971, 422). One simply cannot read him like most literature, and therein lies the difficulty. Metaphysical concepts, mathematical constructs, religious beliefs, allusions to myth and even cosmologies, may enter the picture in one sentence only to be, apparently felicitously, impugned a few sentences later. As a result, the reader must fill in spacious gaps in order to appreciate the connections being suggested. That is to say, one must supply a particular reading, for the text can be given a multiplicity of readings—an activity which, of course, accords with leading critical practices today.

Over the long haul, few writers have been as skeptical as Borges concerning the value of ideas, and even of literature, in

spite of the fact that he says he is "infested with literature." No one, he believes, can lay claim to originality. The sum total of all metaphysical doctrines represents a system of mutually negating entities, each having its antithesis somewhere. Nevertheless, Borges, like many of his characters, defiantly plays a game between the ideal world of the limited intellect and the "real world," which mocks the intellect's every effort to understand it. One might say that Borges's only personal metaphysics is *paradox*. It is to be expected, consequently, that the thinkers who most captivate him tend to dwell on the bizarre, on uncanny twists of the mind. These thinkers are generally not the most noteworthy in Western history. They include Zeno instead of Plato, Berkeley instead of Hobbes, John Wilkins instead of Locke, Jung instead of Freud, Pascal instead of Descartes, and writers such as Kipling, Stevenson, H. G. Wells, de Quincey, and Chesterton. Even among his compatriots, Borges prefers the rather obscure Macedonio Fernández to better-known philosophers such as José Ingenieros. Along these same lines, it bears mentioning, in 1924 Borges as a young man perceived the Spanish poet Luis de Góngora and his followers to be exclusively interested in the formal aspects of verse, while a somewhat lesser-known Spaniard, Francisco Gómez de Quevedo, became a model for the poetry Borges desired to pen: lines in which the thinking process is paramount and intimately linked to the aesthetic dimension of the work (Rodríguez Monegal 1978, 182).

The French intellectuals' fascination with Borges, which began in the early 1950s when Nestor Ibarra and Roger Caillois translated him into French and mushroomed after 1961 when he shared the Formenter Prize with Samuel Beckett, is not surprising. Even though Borges inclines toward Anglo-Saxon thought and literature, there is an affinity between his work and the analytic but elliptical French mind. Michel Foucault confesses that his *Les Mots et les choses* (1966, translated as *The Order of Things*, 1970) was born from a text by Borges, namely the bizarre imaginary Chinese taxonomy found in his essay entitled "The Analytical Language of John Wilkins" (*OI*, 101–5). Borges's attraction to oblique and unexpected conceptual warps and his denial of any and all interpretations are rather compatible with Jacques Derrida, Gilles Deleuze, and the later Roland Barthes. And his fascination with the conjunctive fusion of disparates, especially the archaic with the modern, is also prevalent in Mi-

chel Serres, among others. In short, Borges's prose was tailor-made for the French intellectual climate of the 1960s.

This characteristic of Borges is germane to the thrust of the present inquiry, which is to map connecting lines between Borges's work—especially his metaphysical prose and essays—and certain aspects of twentieth-century mathematics, logic, and physics, and to an extent philosophy of science. I am, of course, in no position to conjecture that these disciplines have "influenced" Borges or "caused" a shift in his mode of thinking. What I do suggest is that Borges, like all writers, participates in an exceedingly complex cultural matrix. Out of this matrix, and revealed in the culturally aware intellectual's work, one finds the product of a general climate of opinion, a "temper of the times," which potentially brings about a convergence of thought and speculation from disparate fields of endeavor. I am not speaking here of any bleary-eyed allusion to *Zeitgeister.* Rather, I submit, leading intellectuals from all walks are sensitive not only to the state of affairs in their chosen field of endeavor but to contiguous fields as well, and they have the skill to incorporate this awareness into their work, at times even to bring about changes. Borges's metaphysical stories and essays are such an imaginative response to the complexities, uncertainties, and ambiguities implicit in many contemporary modes of thought and conduct.[2]

For fear that the reader be given a false impression concerning my objectives, I must be clear on one point. Borges's meditations on Sufi thought and the Kabbalah and his frequent allusions to mysticism might lead one to conclude that I am indirectly suggesting parallels between Borges, mathematics, physics, and Oriental thought, and hence my game would be comparable to much of the so-called "pop physics" that has recently appeared on the scene.[3] It can rightfully be contended that such similarities would "prove" little, for similarities can always be found, or invented, as it were (Goodman 1970). For example, it has been argued that the *I Ching,* with its eightfold extensions, is similar to the "meson octet," in which nuclear particles and antiparticles occupy opposite positions. Jeremy Bernstein (1982, 338) cryptically points out that it "is also vaguely similar to the Ferris wheel in an amusement park." In the present inquiry, I do not engage in a spurious quest for analogies or alternative pathways to truth. Nor do I discount, as caustically as does Bernstein, the suggestive though occasionally

strained links between nuclear physics and Oriental thought. In general, I believe these studies provide interesting and provocative reading, vaguely disclosing some possible directions that the "new physics" may be taking.

Most appropriately, it might be said that the connections I attempt to disclose are the product of "intertextuality," an "intertextuality," as I shall argue, that reaches beyond the limits of literature and branches out into philosphy and even the "hard" disciplines. This inquiry, then, is a dip into the continuous stream of intellectual and artistic endeavor in an attempt to demonstrate how multiple textual threads are intricately intertwined.[4]

This approach, I submit, is justifiable, since traditional lines of demarcation between disciplines are rapidly converging: physics with chemistry and biology, biology with psychology, psychology with linguistics, history and philosophy with critical theory, and so on. This is not mere coincidence. A revised world perspective, slowly emerging since the latter part of the nineteenth century, has been catalyzed by relativity and quantum theory, modern and postmodern art, theory and criticism, and departures from conventional metaphysics. This perspective cannot be adequately appraised, I submit, from within a relatively confining frame of reference. It follows, as a necessary corollary, that the broadest possible view of Borges's work, or of the work of any other writer of comparable stature, is compulsory.

Although this Preface and certain chapters are concerned primarily with Borges, much of the book does not always focus directly *on* his work. More appropriately, Borges constitutes the hub whose spokes point to a constellation of mathematical, logical, scientific, and philosophical concepts of the present century. I have not intended to write a comprehensive study of the Argentine man of letters. Adding yet another critical work to the prodigious heap would be somewhat pointless. Rather, from a postmodern perspective, I use some of Borges's writings as a springboard in order to relate them to multiple issues.[5] If this study appears at times overburdened with references to inquiries outside the humanities, it is not because I wish to overwhelm the reader or evince pretenses to an erudition I do not possess. Rather, I rely on the work of intellects in fields other than my own clearly because I lack their knowledge and competence. Neither do I present the ideas and opinions of others chiefly for the sake of criticism or counter-

Preface argument. My objective—to illustrate commonalities found in disciplines that are customarily assumed to be discordant—is considerably more modest. Given this objective, I largely offer unexpected juxtapositions rather than critical encounters, suggestive parallels rather than exposition or description, synthesis rather than analysis.[6]

In short, each chapter in this inquiry gyrates around Borges, with direct reference to broad themes prevalent in our times. Chapter One actualizes the age-old controversy between realism and nominalism about which Borges has a few intriguing comments. Then I turn to an issue that has taken on increased prominence during this century: the interface between fiction and the "real." In Chapter Two, I introduce another topic which has resurfaced in contemporary times: our ubiquitous paradoxes, especially those of Zeno, which never ceased to captivate Borges. Chapter Three outlines the inevitability of paradox in the purest of human thought, mathematics, and logic. In addition, the problems of randomness and order, the infinite and the finite, continuity and discontinuity, the One and the Many, lead to a brief disquisiton, to be fleshed out in more detail later in this study, of the intriguing phenomenon of consciousness and its relevance to the quantum theoretical view. An Interlude that follows illustrates various affinities between scientific and literary imagination. Then, in Chapter Four, the Einsteinian universe is introduced and related to Borges's notorious Library of Babel. The perennially disconcerting enigma of time, in view of current trends in relativist cosmology, constitutes the essence of Chapter Five, followed by a chapter introducing the perplexities of quantum theory, its promises and disappointments, with the destruction of cherished beliefs and the rupture of desired symmetries. Finally, I expatiate, in Chapter Seven, on the problems of natural language as a communicative medium, an issue that contemporary scientists and writers alike share. What I offer, hopefully, is a contribution to our incessant pursuit of that key enabling us better to comprehend, ever-so-slightly, the nature of our relationship with the world, that is, the world of "textuality."

Acknowledgments

I wish to acknowledge my appreciation to Purdue University for granting me a fellowship with the Center for Humanistic Studies, during which time I wrote the initial draft of this book. In general, I owe inestimable debts to ideas and inquiries of other scholars whose number is so great that its listing would be impossible here—or perhaps, the references might be construed as such a list. More particularly, I wish to thank my colleagues Paul Dixon for the title of this book and Virgil Lokke for his comments on Chapter Six. Permission to publish a small Portion of Chapter Two was granted by the editor of *Language and Style,* and parts of Chapters Four and Five by the editor of *Latin American Literary Review,* for which I remain grateful. I must also confess that, if over the past decade I had not been able occasionally to trap a handful of students in my classroom and air out a few vague intuitions, assumptions, ideas, and doubts, as well as tune in to their frequent penetrating comments and rebuttals, which marvelously served to buffer my vanities, this book would surely have suffered certain impoverishment. And, finally, my debt to Araceli—for her perseverance, her integrity, her seeing me through the thick and thin of things—is such that without it the following words would never have seen the light of day.

Chapter One **Intellection and Contemplation: An Impossible *Coniunctionis Oppositorum***

Much of Borges's most metaphysical work bears directly on the ageless nominalist-realist controversy. In this chapter, I first illustrate that Borges's notion of our being inveterate nominalists is partly vindicated by contemporary science and the computer revolution. This leads to a brief inquiry into the nature of Borges's fictions—the relation between Borges and the rather positivist "as if" nominalism of Hans Vaihinger, and Alexis Meinong's strange and bloated world of "mental objects." Literary fictions, qualified as alternative worlds, or extrapolations from the world as we know it, will be placed in a broad context that includes the myriad fictive constructs sprinkled liberally throughout mathematical proofs and scientific treatises. It will become evident that there is a contradiction here: the fiction *is*, or *says*, what the world at least in part *is not*.[1] Like an oxymoron, it both negates and affirms; it manifests something ordinarily perceived to be intelligible and something else which remains at least partly unintelligible from a literal viewpoint. Understanding a fiction, in this respect, calls for identification of something invariant, that which is true of both the fiction and the "real"—in the midst of difference—and that which is inevitably an "irreal" part of the fiction.

Chapter One

The deeper you try to go into the character of these universal relations which have always been the subject of philosophy, the less you feel inclined to make any pronouncement about them whatever; because you become ever more aware how unclear, inappropriate, inaccurate and one-sided every pronouncement must be.
—Erwin Schrödinger

Metaphysics of Deceit

1 Borges occasionally alludes to Samuel Coleridge's observation that we are all born either Platonists or Aristotelians (*OI*, 56; 123; *L*, 146).² The Platonists believe the universe to be a vast, orderly cosmos, the Aristotelians that whatever we believe the universe to be, it is ultimately the product of our constructive imagination. The former long for the security of their eternal archetypes; the latter deny the existence of archetypes altogether, preferring only to indulge in minutiae of the universe that are close at hand (*HE*, 34). Throughout the ages these indefatigable antagonists have never ceased to lock horns; only their names and language have changed. On the one side are the realists: Spinoza, Kant, Francis Bradley; facing them are the Aristotelians, embracing nominalism: Locke, Hume, William James (*OI*, 156). Borges proceeds to the valorous conclusion that nominalism, "which was formerly the novelty of a few, encompasses everyone today; its victory is so vast and fundamental that its name is unnecessary. No one says that he is a nominalist, because nobody is anything else" (*OI*, 157). It appears strange that Borges would choose summarily to spurn realism while elevating nominalism to the status of a generality. Is Borges himself truly a nominalist, or is this merely another of his strategems of deceit?

Jean Wahl (1964) sees nominalism and realism in Borges as a contradiction. It must be stated in Borges's defense, however, that he never implies he simultaneously adopts both; more adequately stated, he fuses them to create uncanny conjunctions. For example, he observes in "The Nightingale of Keats" (*OI*, 121–24) that the Anglo mentality is incorrigibly nominalist. He then curiously evokes Schopenhauer—from the Germanic tradition, those supreme abstractors—to explain Keats. The poet identifies a particular bird with the class to which it belongs, thus abstracting it from the concrete world and the

experience of time, and so elevating it to the ideal, eternal realm, i.e., nominalism amalgamated with a realism slightly tinged with idealism. This transposition, says John Sturrock (1977, 26), "is achieved simply enough by naming a real thing, by exchanging objects for words." To name something is automatically to fictionalize it, to render it unreal, to replace the thing with a word, to substitute language for the world. Borges, "the archetypal writer of fiction if ever there was one," continues Sturrock, is in this respect a realist, "deeply gratified by the way in which language abstracts from essences to existences." However, Sturrock (1977, 22) claims elsewhere that Borges is not a realist but an idealist. Strangely, there seems to be some truth in this assertion also, given Borges's declaration that "today, were I to choose a single philosopher, I would choose him [Schopenhauer]. If the riddle of the universe can be stated in words, I think these would be his writings" (*A*, 216–17).

On the other hand, Jaime Rest (1976, 49–63) argues that Borges's thought stringently follows nominalist lines. He even hints that Anglo-American empiricism and, surprisingly, positivism, govern much of Borges's reasoning. Though at the outset this assertion might appear outlandish, one must admit that it gives food for thought. Indeed, it compels one to look further, to the meaning-in-use philosophy of the later Wittgenstein (1953), to Hans Vaihinger's (1924) own brand of positivism, his philosophy of "as if" and its relation to Borges—briefly elaborated by Carter Wheelock (1969)[3]—and to Bertrand Russell (1926), whose early writings Borges knew well. But for the present, questions remain. Borges, of course, creates fictive worlds by the appropriation of linguistic signs. Are these multiple worlds standing precariously on the formative function of signs in any sense "real"? If so, do their contrasting features interact, thus constituting larger portions of one totalizing world? Is the one nothing more than a collection of the many? Or, on the contrary, are fictive worlds concocted by the mere manipulation of indefinitely permutable signs?

Perhaps one of the most forthright nominalists, Nelson Goodman, can provide a clue to Borges's aesthetic inclinations and thought. Goodman declares that his *Ways of Worldmaking* (1978) is "at odds with rationalism and empiricism alike, with materialism and idealism and dualism, with essentialism and existentialism, with mechanism and vitalism, with mysticism and scientism, and with most other ardent doctrines" (1978, x). Then what focus does his book take? It is the product, Goodman tells

us, of Kant's supplanting the structure of the world with the structure of the mind, which was finally replaced by the structure of various symbol systems in the sciences, philosophy, the arts, and everyday discourse. This movement "is from unique truth and a world fixed and found to a diversity of right and even conflicting versions of worlds in the making" (1978, x). There is not one world but a plurality of competing worlds. Or better, if there is one world, it includes a multiplicity of conflicting perspectives, and if these perspectives entail many worlds, then the collection of them is one.

World-making is actually a remaking, we are told. There is no perception without conception; the innocent eye is no more than myth. Goodman (1978, 6) suggests that if we follow the long-standing tradition from Berkeley and Kant to Ernst Cassirer, art historian E. H. Gombrich, and psychologists Jerome Bruner and Jean Piaget, we learn that although "conception without perception is merely *empty,* perception without conception is *dead* (totally inoperative)."[4]

Goodman's world-making, however, is not the equivalent of Borgesian fiction-making. The first entails alternative systems of description of "reality"—in whatever form it takes for the describer—from within distinct frames of reference. World-making must begin with worlds already there. Yet fictions are comparable to Goodman's worlds insofar as they are invented more than discovered, by recognizing new connections, patterns, and structures. This is very much an act of mind, an imposition on that which is perceived and which is in the process invariably altered. Indeed, the world is remade by imposing an alternative world on "reality," and each and every such alternative world is governed, at the outset, by fictive standards. For example, two incommensurable propositions, such as "The sun rotates around the earth" and "The earth rotates around the sun," represent two conflicting perspectives. Do both refer to the same world, or is one merely "true" from one frame of reference and the other "false" (i.e., fictive) from another? If the latter is the case—as Goodman believes it is—both can be (and have been) considered "true." Then must there not be a multiplicity of actual worlds? Of course, says Goodman. But each of these worlds would be tantamount to a fictive construct, taken *as if* it were "real" from a particular perspective. In like manner, Borges's fictions, indeed, all literary fictions, are hypostats accompanied by a tacit but cordial invitation that the reader properly construe them *as if* they were "real" from the frame of reference Borges provides.

But a contrast of the Ptolemaic and the Copernican worlds is insufficient for the present discussion, since they both stem from scientific perspectives. A more adequate contrast might be established between a scientific perspective and down to earth, commonsensical knowledge. For example, the physicist Arthur Eddington (1958b, xi–xiii) once referred to his writing desk as a solid object from one view, and as a confusion of largely empty molecules from another. Is it the same desk? Or is it two different desks? Goodman would answer with a firm yes and a firm no to both questions. For the realist there *is* a world, one world. The idealist responds that conflicting frames of reference do not necessarily imply that there *are* many worlds. Goodman (1978, 119) tells us from his nominalist vantage point that both views are "equally delightful and equally deplorable," since the difference between them is no more than conventional. We are free to mark out the frame of reference wherever we like, or we can dance in and out of frames with the facility of the physicist between the particle and wave concepts of the electron, for, aware of the conventionality of all frames, we need not be committed to any frame. Considering, then, that Eddington's two contradictory desks belong to alternative worlds, to two frames of reference, it follows that the equivalent of Goodman's worlds are made and remade much in the order of the fabulist spinning his fictions.

Intellection and Contemplation

Let us turn to a couple of diametrically opposed images in Borges' prose that can further elucidate the realist-nominalist enigma in question. Though Borges occasionally expresses disbelief in Platonic doctrine (*HE,* 20–21), he once confessed to Ronald Christ (1967) that he believed he was an Aristotelian but would like to be a Platonist. Commensurate with this report, Wheelock (1969, 24) regards Borges as "philosophically a fox who longs for the simplicity and certainty of the hedgehog, but cannot bring himself to be one." Of course, the fox is a wily nominalist who slips in and out of language particulars, while the hedgehog is a realist who desires to see everything through the same tinted goggles. Wheelock considers the conflict to be that between "perspectivism" and "universal vision," or between the Zahir and the Aleph, from two of Borges's stories by the same names.

The narrator of "The Zahir" begins matter-of-factly: "In Buenos Aires the Zahir is an ordinary coin worth twenty centavos" (*L,* 156). The coin per se is insignificant. Its function could have been provided by one of any number of objects; we

Chapter One   are told that it has been a tiger, a blind man, an astrolabe, a small compass, a vein in the marble of a mosque in Cordoba. Once the object is chosen for its special purpose, however, it becomes a perspective that potentially reaches out to all perspectives. Potentially, that is, because one who knows not how to avail oneself of the strange powers of the Zahir becomes ensnared by it: one cannot forget it.

This inability to ignore the coin becomes the narrator's plight. After he orders a brandy, and before receiving the Zahir in his change, he deliberates on the oxymoron, a manner of speech best reflecting the function of the small coin. The oxymoron is a word "modified by an epithet which seems to contradict it: thus the Gnostics spoke of dark light, and the alchemists of a black sun" (*L*, 158). In other words, the modifier of the oxymoron says what the word does *not* say. The narrator then recalls that he had only a few moments ago left the wake of Clementina Villar—with whom he had obviously been infatuated, and undoubtedly for this reason he finds himself in his present environment. His memory takes him back to previous instances when he had seen Clementina. He notes, very significantly, with implications pointing toward the Zahir, how "the progress of corruption brings it about that the corpse reassumes its earlier faces" (*L*, 158). From that point onward the protagonist becomes obsessed with the small round object.

In good nominalist fashion, and commensurate with Goodman, the Zahir can arbitrarily be anything that stands for something else. Lying in contiguity with all other objects of the world, it is oxymoronic insofar as it is *not* that which would ordinarily represent the represented, and it is capable of all possible perspectives, but these perspectives can be no more than sequential in nature. Like language, the Zahir affords successive perceptual grasps of fragments of the universe, but none of them can be all-embracing. Hence, given human finitude, the incapacity to hold more than a few items of thought mentally in check for more than a fleeting instant ensues: the Zahir is ultimately a helpless symbol. The narrator, realizing this, ends with a futile hope: "In order to lose themselves, in God, the Sufis recite their own names, or the ninety-nine divine names, until they become meaningless. I long to travel that path. Perhaps I shall conclude by wearing away the Zahir simply through thinking of it again and again. Perhaps behind the coin I shall find God" (*L*, 164).

Insofar as a word, like the Zahir, supposedly stands for something in the world, it is meaningful—or, by extension, for those unfortunates who have been temporarily or irreversibly lured into the lair of "word magic," the word *is* the world. And if meaningful, the word can be conjured up and thought about in the absence of that for which it stands. Now let us reverse this formula. If something is thought about long enough and hard enough, it can be, so to speak, conjured into existence, and if so, it can equally be thought out of the world. For Goodman, we find what we are prepared to find or expect to find in the world; thus we conveniently tend to blind ourselves to what is inessential for our pursuits. Tight systematization of our perspective of the world—the world we are incessantly making and remaking—can never be ultimate, for "there is no more a unique world of worlds than there is a unique world" (Goodman 1978, 17). Likewise, the Zahir, whether considering its "thingness" or its "wordness," is a perspective in flux: differentiation through sameness, or vice versa. Perhaps rapt attention to the Zahir's "wordness" can eventually render it meaningless, or perhaps mention of its "thingness" can finally annihilate it from the world—only for it to pop up in yet another world. In any event, the Zahir, as a nominalist key to many worlds, is perpetually condemned to incompleteness. Hence its attraction and its danger.

Borges's choice of a coin to exercise such an overpowering influence is not an isolated case. The narrator of "The Disk" in *The Book of Sand* (113–16), comes upon a Saxon coin, Odin's disk, which, unlike the Zahir, has only one side. Overcome with greed, the narrator kills the traveler who allowed him to see it, and after disposing of the body, he returns to his hut, but try as he may, he cannot locate the disk. This coin, commensurate with a topic to be discussed in the second section of Chapter Three, is a metaphor for infinity, and as such the coin can have no reverse side, for since infinity according to one particular view is incompletable, the reverse side must therefore be invisible.[5]

"The Aleph," like "The Zahir," revolves around the death of a loved one: Beatriz Viterbo. Beatriz is the cousin of Carlos Argentino Daneri, a writer of endless but uncreative poetry in which he reportedly enumerates the entire universe. He is able to do so, he claims, for he has seen an Aleph. He tells the narrator (that is, Borges) of the Aleph's existence in the home of his parents and grandparents, explaining that it is a point in

Intellection and Contemplation

Chapter One   space that is interconnected with and contains all points. Borges visits the house, descends the stairway leading down to the basement where it is located, and he experiences it, a sphere about one inch in diameter, whose center is everywhere and whose circumference nowhere, a beginning without terminus, paradoxically both finite and infinite, the "only place on earth where all places are—seen from every angle, each standing clear, without confusion or blending" (*A*, 23). The narrator then petitions the gods that they grant him the appropriate metaphors with which to describe this miraculous vision, but he knows it is ultimately impossible

> for any listing of an endless series is doomed to be infinitesimal. In that single gigantic instant I saw millions of acts both delightful and awful; not one of them amazed me more than the fact that all of them occupied the same point in space, without overlapping or transparency. What my eyes beheld was simultaneous but what I shall now write down will be successive, because language is successive. Nonetheless, I'll try to recollect what I can. (*A*, 26)

The Aleph affords a realist image as opposed to the nominalism of the Zahir. One entails a transcendental revelation, the other a series of significant perceptual grasps, or in a way of speaking, one is synchrony, the other diachrony. One is a superposition of all objects, acts, and events in simultaneity, the other a serial collection of particulars with no determinate links. According to Borges, today we almost instinctively favor nominalism, but, in spite of ourselves, we implacably gravitate toward the opposite pole in an effort to discover "reality" in the eternal forms. And since this search is ultimately futile, we flee back to the secure minutiae of our empirical world.

This uncertainty is actually becoming a sign of our times. Borges tells us, as do certain other contemporary writers, that we have forever lost the sureties of previous epochs. The strange conjunction of mathematics and theology, from Pythagoras to the Middle Ages, then to Spinoza, Leibniz, Descartes, and even Kant, has been an "intimate blending of religion and reasoning, of moral aspiration with local admiration of what is timeless, which . . . distinguishes the intellectualized theology of Europe from the more straightforward mysticism of Asia" (Russell 1946, 37). Borges, somewhat contrary to this Western trend, intellectualizes supremely, especially during the writing

of *Ficciones* (1944), *El Aleph* (1949), *Historia de la eternidad* (1936), and *Otras inquisiciones* (1952). In these writings there is a conspicuous absence of allusions to the traditional Christian God. Even in the few stories directly involving Christ and theology, the argumentation is medieval, scholastic in nature. And in spite of influences from the Sufis, and even the Kabbalah, Borges continues to intellectualize.

Intellection and Contemplation

Perhaps Borges is in essence a realist temperament disguised as a nominalist, then. In this light, judging from his frequent intellectual excursions into mysticism, it should come as no surprise that he often longs for a view *sub specie aeternitatis,* such as that afforded by the Aleph, for example. On the other hand, he has remarked that all perspectives, all classifications of the world, are nothing more than convenient intellections, that is, fictions (*OI,* 104; Charbonnier 1967, 85), and that individuals exist only insofar as they are participants of the species that include them, which is the only valid "reality" (*HE,* 18–19). Borges then reiterates his Schopenhauerean postulate of Keat's nightingale, which, he asserts, is identical to the one heard by Ruth in the fields of Bethlehem, and he goes on to tell us that, like the nightingale, every nonhuman organism, unaware of death and time, exists in the eternal now. All are identical with themselves and with the species. To *be,* in the eternal now, is to *exist* platonically. We have, in a certain way unfortunately, been banished from that felicitous Eden, however. The eternal is no longer available to us; nonetheless, in Borges, the longing lingers.

It is becoming apparent that Borges, like Goodman, takes metaphysical doctrines, scientific theories, and other particular perspectives of the world at face value. Each is *a* world, none is *The World*. All worlds depend upon a unique vantage point: the earth as center or the sun as center, the universe as finite or infinite, the electron as particle or wave. And the conjunction of all possible perspectives makes up a *mutually cancellatory* collection providing fodder for Borgesian fictions. Such fictions are good to think with, of course. They also illustrate that any and all perspectives of the world, rather than resting on solid foundations, are based on subjective, though sometimes rigorous, premises that could always have been other than what they are. Hence Borges's skepticism of all doctrines, compelling him toward what might appear to be random world-making, is not a concession to gratuitous esoterism or sophistry. It is, more adequately, the most enlightened of gnoseological postures.

Chapter One

Although Argentine and, one might suppose, outside the mainstream of continental European thought, Borges is not as lost in metaphysics as he often claims (Alazraki 1971, 23).

In fact, Borges's writing is contemporaneous with an emerging world image in Western technological societies. Although he appears to be neither exclusively a nominalist nor a realist, the fact remains that he claims we all are, in spite of ourselves, nominalists. This pretense is commensurate with the conclusion of David Bolter (1984, 132–38) that we have entered the age of "computer thought," which began during and shortly after World War II and was accompanied by the development of information theory and cybernetics. Signs manipulated by computer thought are discrete, conventional, and finite. This is, relatively speaking, a recent development. The ancient Greeks were mesmerized by the resonant qualities of the spoken word. The word and the idea it incorporated were conceived to be as real as the chair I am sitting in. The sophists continued to so use the spoken word, but, rhetoricians that they were, they taught their disciples to regard language as something manipulated arbitrarily for the sheer purpose of suiting their fancy. Language, in their eyes, had lost much of its awesome power. Plato, of course, had no use for this radical new article of faith. He could not regard any facet of the world to be arbitrary. The written sign became the culprit, for he believed it to possess an abstract, distancing power threatening to relegate language to the status of whimsical marks and scratches, unembodied with determinate meaning. In spite of the birth of modern science and the predominance of writing, the Platonists, in a manner of speaking, had their way for generations. The sound of language was retained in ancient reading, for texts were to be read aloud to be adequately intelligible—in much the sense that one must play a musical score to properly appreciate it. In medieval times the debate concerning the relationship between words and their referents became the realist-nominalist controversy—found at the core of Borges's "metaphysical" narrative. But the magic of the spoken word continued to reign supreme, at least until the age of the printed word, when the visual linguistic image gradually became the medium for assimilating knowledge.

Mathematics followed the same route. The Pythagoreans equated the cosmos with number, and number with a mystical conception of geometry. Their mathematics remained chiefly verbal; axioms, definitions, lemmas, and proofs were articulated in natural language so that they might be intelligible to anyone. Even down to the time of Galileo, mathematics re-

mained tied to the spoken word. In fact, Galileo himself declared the book of nature to be written in the language of mathematics, but it was a book to be read aloud, for understanding was for him auditory, not visual. With the introduction of analytic geometry and calculus, arbitrary symbols became the order of the day. Mathematics was now transformed into a language in its own right, an artificial language. And significantly, Leibniz's binary system based on the root two paved the way for modern computer language. The result has been more astounding than the wildest of dreams. From the Western tradition of realism according to which the mind discovers ideas that have a power, necessity, and reality of their own, modern thought, following the computer revolution, entails invention rather than discovery, which suggests no necessary link between thought and the world. "Computer thought," then, very effectively represents the triumph of nominalism, and William of Occam appears to be having his heyday (Bolter 1984, 77).

Today's nominalists, of course, claim that the only thing so-called universal classes of entities have in common is their name, and that this premise is closer to the "truth" than the realists' notion that words are somehow embodied in the furniture of the world. Yet, viewed from a broad historical perspective, and especially in light of contemporary physics, as we shall note, one must admit that mathematical language might eventually turn the advantage over to realism. If natural language cannot do justice to the "reality" of Eddington's desk, mathematical symbols, it is now believed, can relatively adequately describe it. In other words, the cleverly manipulated mathematical symbols on the page *become* the "reality." The problem is that these symbols do not incorporate things in the world, for they enjoy no determinable empirical counterpart.

Let us, then, speak briefly of mathematics and its relation to "reality," if indeed there be any, for the purpose of comparison with Borges's "nominalistic" fiction-making by means of intellection. Bertrand Russell, in *The History of Western Philosophy* (1946)—with which Borges was quite familiar—juxtaposes a pair of antipodal terms, "mathematical" and "empirical." This is fundamentally Borges's conflict between realism and nominalism, which, I believe, is clarified somewhat in the context of contemporary science (and science has in a roundabout way become the issue here.)

Availing himself of Russell's dichotomy, Albert Einstein, one of the most philosophical of twentieth-century physicists,

Chapter One argues that scientific concepts have two sources of meaning (see Northrop 1949). One is empirical, providing concepts from particulars. The other is formal, expressed in mathematical symbols capable of generating universal concepts related to "reality," since they derive their meaning by postulation from general propositions. "Nominalism" in science entails operationalism, utilitarianism, that is, what the practicing scientist actually does—his pencil and paper calculations, manipulation of instruments, scanning of computer printouts, recording of meter readings, prediction of future readings—with little immediate regard for "reality out there." The "realist," in contrast, constructs hypothetico-deductive theories, the mathematical terms of which he believes represent actual objects making up the world. Both of these activities are essential to science. If the nominalist were to maintain exclusive rights, there would hardly be any scenery to photograph and hence no photograph. If realism were to exercise a monopoly, the photograph would be no more than metaphorical; it could be validated solely by comparing it to some hypothesized "real world." Yet without some acquiescence to realism, neither knowledge nor ignorance would be forthcoming. There would be only varying degrees of success or failure without the necessary assumption of something "out there" about which abstract descriptive statements might be formulated: science as we know it could not exist (Clarke 1977, 111–17).

In this light, interestingly enough, Raul Gutiérrez Girardot (1959, 58–60) perceives two types of language used by Borges: linear language (of logic and mathematics) and intuitive language (of images). Roughly, the first is to the theoretical language of physics as the second is to operational (nominalistic) language. The former entails construction of paradoxes, infinite regresses, geometrical structures, symmetries, and labyrinths; the latter reveals the aesthetic exploitations of the former. This is, of course, an impossible conjunction. Intellection can only be contradictorily wedded to intuition or contemplation. Nevertheless, in view of Eddington's "schizophrenic" desk, Borges's incessant flitting between the two horns of this dilemma effectively mirrors twentieth-century humankind's efforts to know the world.

Actually, it befits Borges that he engage in a balancing act between ideas and doctrines. He finds mischievous delight in playing around with ideas purely for the imaginative possi-

bilities they offer, and in this process the metaphysics, logic, and mathematics in his prose invariably suffer a morphological transformation. Indeed, Borges's fictions are "hyperfictionalized" in such a manner that their underlying conceptual substrate surfaces without the fiction suffering any impoverishment. To cite some brief examples, Anglo empiricism of the eighteenth century is extrapolated to the extreme in the mysterious planet Tlön of "Tlön, Uqbar, Orbis Tertius" (*L,* 3–18). Kurd Lasswitz's concept of a Universal Library becomes "the Library of Babel" (*L,* 51–58), with its myriad hexagonal galleries, its innumerable books containing all possible combinations of a finite repertoire of symbols, and its pathetic inhabitants overwhelmed by the magnitude of this monstrous athenaeum.[6] The extracted essence of Hume and Berkeley unites strangely to negate time in "New Refutation of Time" (*OI,* 171–87). J. W. Dunne's bizarre study of consciousness and dreams is eulogized, yet ironically disbelieved, in "Time and J. W. Dunne" (*OI,* 18–21). In "Three Versions of Judas" (*L,* 95–100), the Christian concept of betrayal and sacrifice is taken to its logical extreme to create an apparent absurdity. And idealist philosophy in a sense backfires in "The Circular Ruins" (*L,* 45–50), where a magician who has dreamt a son and finally succeeds in interpolating him into "reality" discovers in the end that he too is a dreamt image of yet another dreamer. And so on.

Borges realizes that the word remains inextricably against the world, that there is no way to divorce oneself totally from metaphysics, or from language. The world, to reevoke Goodman, is a plurality. The impossibility of penetrating the divine plan of the universe, nonetheless, does not discourage Borges from inventing limited human constructs, always remaining mindful that they can be no more than provisional. Jaime Alazraki (1971, 27–28) rightly points out that

> [l]ike Don Quixote, Borges does not read reality as it is (a task for the gods, not man). On the contrary, he follows the signs coined by culture. Unlike Don Quixote, who desires to transform reality into a sign in order to demonstrate that books reveal truth, Borges suggests in his story ["Tlön"] that all of reality has already been transformed into a sign, and that this symmetrical sign is now as far removed from reality as were the novels of chivalry in Don Quixote's Spain.

Chapter One

Signs (books) are all we have, for better or for worse: the receding paradise of "word magic" as well as Platonic doctrine appear to be elegant but futile dreams. And computer "thought," with its nominalistic trappings derived from digital, linear, binary logic, threatens to engulf us, just as in Borges's story our world gradually becomes mere idea, that is, the imaginary Tlön. Of course, I am speaking of no tidal wave that will carry us out to sea in one fell swoop. The change has crept, is creeping, up on us. In the nineteenth century, Boolean algebra and the beginnings of modern logic initiated by Charles S. Peirce, Augustus de Morgan, Giuseppe Peano, and others evolved into Gottlob Frege's foundations of arithmetic, Russell and Whitehead's *Principia Mathematica,* symbolic logic, and artificial computer languages. William Rowe Hamilton's algebra of "quaternions," using imaginary numbers (i.e., $\sqrt{-1}$), Georg F. B. Reimann's non-Euclidean geometry, and Felix Klein's theory of groups have been conveniently appropriated by physicists to aid in describing Eddington's vacuous desk of swarming electrons. Yet, accompanying this trend is the increasing awareness that the "truth" of scientific theories cannot be guaranteed. The sign, whether linguistic or mathematical, is not, and cannot be, coequal with the world.

What generally goes unacknowledged, however, is the fact that certain eighteenth-century philosophers, those who exercised the most profound influence on Borges, were saying somewhat the same thing—albeit in a different language—and hence, as we shall observe below, the Argentine's thought is strangely antiquated and at the same time contemporary. For Hume, our complex ideas are nothing more than a collection of simple ideas, memories, images, and thoughts, and we are no more than a bundle of sensations. Furthermore, our mind is not a distinct entity in its own right; it is identical with these sensations. This being the case, all we can know are our sensations of the world "out there." Since we cannot directly know the world, repeated perceptual grasps of Eddington's desk do not prove its existence. We cannot even say it exists in space and through time, since space and time are merely the order in which our sensations occur. The inference of an external world, then, enjoys no necessary validity, and scientific laws governing this world are equally invalid. The mathematician Morris Kline (1980, 75) concludes that by "destroying the doctrine of an external world following fixed mathematical laws, Hume had de-

stroyed the value of a logical deductive structure which represented reality." Berkeley, who had an even greater impact on Borges, denied the absolute existence of anything beyond our perception of it. Thus the writing desk *is* only if it is perceived. In a sense Berkeley's thought was less radical than Hume's, who was so contrary as to be absolutely repugnant to many intellectuals of the eighteenth century. There is no invisible, intangible object called matter outside our sensory impressions of it, so the world does not exist as such; only we, or an omniscient God, can call it into "existence."

Very significantly, in view of these two doctrines, Borges writes in "The Immortal" (*L,* 105–18), a story about a race of Troglydites who have tasted of the river of eternal life: "We accept reality easily, perhaps because we intuit that nothing is real" (*L,* 113). Since the Troglydites' life will be perpetuated forever, they will eventually experience all things, each one of them will be every one (in the sense of Schopenhauer), there will be no novelty, no surprises; expectations will cease, or perhaps there will be the expectation merely of nothing. The Troglydites have come to live almost exclusively in their thoughts. They hardly perceive the physical world, for the world has become mere idea (or will), and they have become mental scanners, cerebral rats; their inner thoughts *are* their world, their signs are their only "reality." In other words, they have become counterparts to contemporary physicists—supreme mathematicians—whose sign constructs depict a nonempirical, unimaginable realm: Eddington, his desk, the pen in his hand, the paper before him, are not really "real." What is "real" is what common sense would ordinarily dictate to be the "irreality" of it all.

In the manner of this century's most adept abstractors, it will become increasingly evident throughout this inquiry that Borges, master of deception that he is, "hyperfictionalizes" to the extreme in his stories and essays, encapsulating the distilled essence of the metaphysics underlying the most rigorous contemporary thought. During such "hyperfictionalization," the fox, without warning, may provisionally become a hedgehog, or vice versa. Or, both may temporarily disappear altogether between the interstices of Borges's mind, awaiting their appropriate recall, which, of course, depends upon their master's aesthetic whims. In order to get a preliminary grasp on this "hyperfictionalization," let us journey briefly through Hans Vaihinger's world of "as if" fabrications.

Chapter One

We may observe that, in expressions, the mathematical language has become entirely visual, there is no proper spoken form, so that in reverbalizing it we must encode it in a form suitable for ordinary speech.
—G. Spencer-Brown

Where Fiction Ends

2 Wheelock (1969, 26) remarks that some of Vaihinger's ideas, "taken at random, are so useful in speaking of Borges' idealism and the esthetic, and the attitudes of the two men are so similar," that he cannot resist "using one to place the other in bolder outline." Manuel Ferrer (1971, 69n) much less cautiously asserts that the influence of Vaihinger on Borges is profound. He goes on to ask why Borges, whose citations often become repetitious, and who alludes to numerous dogmas, beliefs, and philosophies throughout his work, becomes strangely silent regarding those who have most influenced him. Borges, of course, is a poet, a fabulist, and an essayist. He need not share the scholar's obsession for acknowledging sources. On the other hand, he is notoriously elusive; he often intentionally leads readers, audiences, and interviewers down blind alleys. We might do well, then, to pursue not reputed influences but affinities between Borges and Vaihinger.

Vaihinger would have us believe that the fictive activity of the mind is the product of fundamental psychic forces. The mind, inventive and under the compulsion of necessity, discovers itself in a world replete with contradictions. Exposed to the assault of this hostile world, to preserve itself it resorts to every possible means, paramount among them being the creation of fictions in an attempt to generate order out of disorder (Vaihinger 1924, 12).

Ben Belitt (1972, 212) suggests that, for Borges, to write stories rather than to tell the truth—which is presumably the enterprise of science, and perhaps metaphysics—is "to dissimulate, and all dissimulation, rationally viewed, confounds, rather than assists, reality." To validate Borges's posture, Belitt shortly thereafter alludes to Vaihinger. Belitt's remark is somewhat misleading. Scientific explanations, as I shall argue below, are like stories in the sense that they are used to render the unintelligible relatively intelligible upon relating the known, by means of a fiction—i.e., a scientific model, or a "thought experiment"—to the unknown (Hesse 1966, 7–56). Scientific theory-making, it

is becoming increasingly—and at times painfully—evident, also entails dissimulation, for since all scientific models to a greater or lesser degree fall short of the mark, they, so to speak, must "tell a lie" (Feyerabend 1975; 1978).

This is, in essence, Vaihinger's thesis. For example, in mathematics, surfaces, lines, and points are fictions, forms without content. They are nothing, but a "nothing that is nevertheless conceived as a something, a something that is already passing over into a nothing" (Vaihinger 1924, 234). A line is the "boundary" of a surface. But what is the surface? A mathematical slice, curved or straight, out of the conception of pure, absolute mathematical space, which is also a fiction. And what is the line? An infinitesimally thin imaginary nothingness separating what is inside from what remains outside. Nevertheless, all these fictions are useful, and even necessary, insofar as they have a bearing on our conception of the world.

Perhaps the most astounding fiction used in mathematics is infinity. To calculate the length of an arc it is assumed to be a sequence of rectilinear contours with an increasing number of sides. As the number of straight lines of this sequence approaches a limit, the length of the arc is defined as the limit of the sequence. And what does this limit, which adapts the straight to the curved, ultimately become? An inconceivable, infinite number of steps. The curved world of our senses is the quasi-final step in an infinite sequence of infinitesimally flat worlds that can exist only as an act of mind. In a rather Vaihingerian vein Alex Comfort (1984, 196) refers to the use of mathematics in describing subatomic particles:

> All mathematization is metaphor: to describe observation, we have a kit of Erector parts (polynomials, spinors, matrices, etc. etc.). If we develop an algebra for the velocity of a wave packet, we can describe its behavior in frequency terms, the "oscillation" is a metaphor: particles possess "spin" but there is no physical rotation going on. In the last resort *metaphor is all we have.*

In general, mathematical fictions, indeed, all fictions, scientific, Borgesian, or otherwise, are characterized by their arbitrary deviation *from* "reality," and especially by the violence they do *to* it. Thought is circuitous, and therein lies one of the secrets of fiction-making, since fictions are "temporary halting-places for thought and have no bearing on reality" (Vaihinger 1924, 100). An even more general characteristic of fictions is

that they *are* in part what the supposedly "real" in part *is not,* i.e., the oxymoronic character of fictions, as outlined above with respect to the Zahir. This introduces the function of *negation* as well as Vaihinger's debt to Nietzsche, for whom we must learn to lie, for it encourages the art of illusion, and we must learn to love illusion, that is error, for it is the mother of all knowledge. In a complementary mode, Umberto Eco (1984, 167–82) positions the lie as the cornerstone of all human semiotic: without the capacity to construe a sign *as* something which it *is not,* or to use something to denote what it would ordinarily *not* denote, human language as we know it could not have blossomed into existence.[7] Eco's notion harks back to the nominalism of the medieval period, according to which all general ideas are *ficta, fictiones,* without there being attached any positive meaning to the fiction. Vaihinger, on the other hand, prefers a less "negative" sense of fiction insofar as one must realize that the fiction, though a fiction, can nevertheless serve one in acquiring knowledge—albeit indirectly.

Vaihinger, it must be admitted, does not meet the approval of the entire mathematical community. Hugh Lehmann (1979, 66–70), among others, rejects Vaihinger's approach because it accepts a mathematical formulation provisionally as true, though it might be self-contradictory and/or inconsistent. Contradictions and inconsistencies do not inconvenience Borges or Vaihinger, however, for they assume inconsistencies to be a sine qua non for good fictions. This view also coheres appropriately with Nicholas Rescher and Robert Brandom's (1979, 31–33) "logic of inconsistency," which differentiates between the "logically possible" and the "semantically possible." The first must be logically consistent and compatible with Aristotelian principles; the second need be only "describable," "conceivable," or "thinkable." Fictions, in the Borgesian (and Vaihingerian) sense, are clearly the source of nonstandard semantically possible objects, which are often contradictory or inconsistent, in addition to their being indeterminate and incomplete—such as Hamlet's hat size, his shoe size, the number of hairs on his head, whether or not he has a mole on his left leg, etc. (Woods 1974). The question obviously arises, however, on using the term "semantically possible *objects*": If something is thinkable and sayable but cannot exist, i.e., a square circle, is it an "object"? For that matter, even certain mathematical concepts, especially $\sqrt{-1}$—that amphibian between being and nonbeing, as Leibniz called it—pertain to this category of thinkable and

sayable but nonexistent "objects." After a brief excursion into an intriguing hypothesis concerning the nature of mathematics, it will be possible to judge more appropriately the import of Borges's, and by extension Vaihinger's, conception of fictions, and of those so-called nonexistent "objects."

Oswald Spengler devotes an entire chapter in *The Decline of the West* (1932)—which Borges cites occasionally—to the meaning of numbers. He concludes, surprisingly—and contrary to Frege and others of his day who believed mathematics to constitute the bedrock of certain knowledge, fixed for all time—that mathematics is not universal. There are different "number worlds," the character of a particular arithmetic depending chiefly on the culture and the conventions in which it is embedded. At the outset this appears absurd. We ordinarily consider counting to be an intuitive capacity: $2 + 2$ cannot by any stretch of the imagination be anything but 4, and our decimal system seems so natural, for we have ten fingers and ten toes. But Wittgenstein, who knew Spengler's work well, has also argued effectively in *Remarks on the Foundations of Mathematics* (1956) that our mathematics need not be the only possible mathematics. In support of Wittgenstein, sociologist David Bloor (1976, 95–116) presents various alternative forms of mathematical thought, concluding that "by exhibiting divergences of style, meaning, association and standards of cogency [it becomes] clear that there are significant variations in mathematical thought which need explaining. Further, it is plausible to suppose that these variations may be illuminated by looking for social causes" (Bloor 1976, 114).

And in a comparable vein, the philosopher of science Stephen Toulmin (1972, 251–52) remarks:

> Such fundamental mathematical concepts as "validity" and "rigour", "elegance", "proof", and "mathematical necessity", undergo the same sea-changes as their scientific counterparts. . . . The result is that the concepts, methods, and intellectual ideals of mathematics are no more exempt from "the ravage of history"—as Descartes and Frege hoped and supposed—than those of any other intellectual discipline.

This notion goes against the grain of mathematical realism, according to which mathematical "objects" exist outside us to be discovered or observed. If the realist were asked why he concludes that $2 + 2 = 4$, he would respond that it is

Chapter One

because 2 + 2 *does* equal 4, for it has been observed to be true. In contrast, Wittgenstein consistently argues against the idea that standards exist independently of human practices. For example, consider continuation of the number sequence 2, 4, 6, 8, . . . For a mathematical realist, the continuation always already exists. One need merely trace it out, and to do so is to move toward the *truth* with each step. After all, the realist continues, how else could there be universal agreement on the proper rule for continuing the sequence?

Wittgenstein, on the other hand, contends that we generally agree as a result of how we were trained in mathematics. As with any other language game, the game of mathematics is a socialization process. There are right and wrong ways to play the game, and the successful player learns to minimize the mistakes:

> *How do I know* that in working out the series "+2" I must write "20004, 20006" and not "20004, 20008"? . . . And if I know *in advance,* what use is this knowledge to me later on? I mean: How do I know what to do with this earlier knowledge when the step actually has to be taken?
>
> "But do you mean to say that the expression '+2' leaves you in doubt what you are to do e.g. after 2004?"—No; I answer "2006" without hesitation. But just for that reason it is superfluous to suppose that this was determined earlier on. My having no doubt in face of the question does *not* mean that it was answered in advance. (Wittgenstein 1956, I: 2)

In sum, from this perspective mathematics is invention rather than discovery, a human institution rather than an eternal playground for the gods. Right and wrong mathematical behavior lies in the game, for mathematics is normative. Consequently, number systems need not be decimal, as is our Western system. To cite only two examples from among many, the ancient Mayas used a system to the base five, for reasons unknown. And within a fictive context, the inhabitants of the planet Tlön use a duodecimal system, which is indeed significant, since Herbert Ashe, one of the "demiurges" responsible for the creation of this imaginary planet, is at one point in the story engaged in converting decimals to duodecimals; as decimals become duodecimals, so the world becomes Tlön.[8] Mathematics, in Borges's tales, is accordingly fictive through and through, much in the sense of Vaihinger, and it is invention, in accord with Wittgenstein. An affinity between literary fiction-

making and the use of mathematical fictions might possibly be validated, then, since both are based on a fictional standard.

If we assume mathematical language to be conventional, then nonmathematical language must be even more so. This being the case, natural language cannot by any stretch of the imagination be the standard bearer for "reality." For instance, Wittgenstein (1972, 14–24) asks how one can satisfy oneself concerning the existence or nonexistence of unicorns. Of course, up to this moment we have found them only in myth, literature, art, and other fictive media. But this is not sufficient. Although unicorns have exclusively populated fictive worlds, we cannot declare with absolute certainty that they *do not* exist in the "real world." What I refer to here is part and parcel of the inductive fallacy. The proposition "All ravens are black" is appropriate to past and present observations, but not to future ones, for, just as I cannot predict absolutely that the sun will rise tomorrow, neither can I be sure that in the future a nonblack raven *will not* pop up. Returning to Wittgenstein's aggravating question, How can we know if unicorns do not exist? the answer can only be, We cannot. So until we experience the instantiation of a unicorn in the "real world," what we can know about unicorns is limited to fictive contexts.

And, within fictive contexts, past, present, and future, our knowledge of unicorns, and of any and all fictive objects for that matter, has no upper bound. For example, in "The Unicorn" (*BI*, 228), Borges reports that according to Ctesias, "among the kingdoms of India there were very swift wild asses with white coats, purple heads, blue eyes, and in the middle of their foreheads a pointed horn whose base was white, whose top was red, and whose middle was black." Borges takes pains to point out that other descriptions of unicorns are less imaginative. As a matter of fact, there is potentially an indefinite number of possible unicorn descriptions, all of them at least partly contradictory. It must be admitted, then, in light of Ctesias's extraordinary unicorn description, that any and all unicorns, proportionate with Rescher and Brandom's "semantically possible objects," are invariably indeterminate and incomplete. They are, properly speaking, fabrications of the mind, and hence can take on any number of forms, depending upon cultural conventions and contexts. To reiterate an above statement about such mental fabrications, all unicorn models, like all other fictions, are invariably in part what the "real world" *is,* and, oxymoronically, what it *is not.*

Chapter One

On the other side of the coin, Borges comments on John Wilkin's artificial language, an elaborate classificatory system consisting of words that "define themselves." He then offers his well-known Chinese alternative to our Western animal taxonomy. All beasts are divided into:

> (a) those that belong to the Emperor, (b) embalmed ones, (c) those that are trained, (d) suckling pigs, (e) mermaids, (f) fabulous ones, (g) stray dogs, (h) those that are included in this classification, (i) those that tremble as if they were mad, (j) innumerable ones, (k) those drawn with a very fine camel's hair brush, (l) others, (m) those that have just broken a flower vase, (n) those that resemble flies from a distance. (*OI,* 103)

Michel Foucault's (1970, xv) remarks on this passage are notorious, and for good reason. The point is that, in line with much contemporary thought on language (1) there is potentially an infinite array of taxonomies of the "real world"; (2) each and every particular taxonomy is, like fictions, incomplete and arbitrary because *words can do no more than define themselves;* and (3) what is deemed a "correct" or "normal" taxonomy is so deemed because it is conventional. In other words, Borges's entry demonstrates that the ultimate taxonomy, implying absolute knowledge, cannot be achieved by a mere classification of particulars. It is possible solely through apprehension of the set of all classifications, and classifications of classifications, etc.—which eventually becomes, for practical purposes, and given human finitude, the equivalent of infinity in mathematics.

Whether speaking of Borgesian fictions or "real world" taxonomies, we are once again reminded of Goodman's world-making. But contrary to world-making, Borges, like any wise fabulist, does not attempt to compete with "reality." He interjects "reality"—properly a conventional "reality"—into his fiction, he does not put fiction into a supposedly verisimilar or mimetic narrative. As such, he constructs a "fictive reality" in which words take their rightful place among other words because words in the text, indeed all words, enjoy no necessary correlation with "reality." Yet one might retort that if, following Vaihinger, mathematics and science consist exclusively of fictions that at least in part coincide with the "real," then it must be possible to pin them down to "reality" at some point. This is not necessarily so, however, for if our knowledge of "reality" is

indeterminately variable, then it cannot be known precisely where the fiction ends and "reality" begins.

Thus, we are caught in a dilemma. Mathematics is not, ipso facto, about "reality." It is, however, appropriated by the physicist to describe an intangible, and for practical purposes, incomprehensible world. But this description can be no more than a Vaihingerian fiction. When one abstract formulation no longer effectively accounts for the "facts," it is to be conveniently discarded and another embraced. So all scientific frameworks, it is becoming increasingly apparent, are merely temporary; and they are conventional.

Borges creates alternative worlds in his metaphysical fictions. But scientists and mathematicians have always exercised the same prerogative. Copernicus's sun is an alternative to Ptolemy's sun, Reimannian geometry an alternative to Euclidean geometry, Einsteinian physics an alternative to Newtonian physics. And so on.

Where, then, is an absolute distinction, if any, to be found between the "real" and the "irreal"? Perhaps, one might surmise, it exists in the distinction between intensional (i.e., the internal world of thoughts, volitions, and emotions, *res cogitans*) and extensional (i.e., the external world of things, *res extensa*) reference. For instance, "Sherlock Holmes lived in London" could be looked upon as properly fictive, intensional, and true, while "London is a city not lived in by Sherlock Holmes" is ordinarily construed to be literal and extensional. So the distinction is elementary. Or is it? According to Goodman's "Earth as the center of the universe" as opposed to "Sun as the center," and a myriad of other comparable examples, what is fictive can become "real" and what was "real" can become fiction.[9]

In light of our intensional worlds, let us introduce the strange case of Alexis Meinong in order further to elucidate the nature of fictionality. In brief, Meinong, following Franz C. Brentano, who claimed that all mental states are directed toward *something* and thus possess distinguishing features, proposed that what *is not* (a property of fictionality) is as important as what *is* (conventional "reality").[10] According to this notion, knowledge not only pertains to existents, that is, to empirical objects, but also to the arts, to imagination, and to all inner experiences. How else, a Meinongian would ask, could theories of the "real world" have come about except by virtue of imaginary worlds? We have no alternative but to conclude, Meinong

tells us, that "objects" that do not, and perhaps cannot, exist, are nonetheless genuine "objects" and part of the total experienced world.

This implies, first, that there are "objects" that do not exist and, second, that nonexistent "objects" can nonetheless be spoken *of* in such a way that they can be made the subject of a predication *as if* they were "real," hence they must be so constituted in some way or other. In this sense there are existent and nonexistent "objects," just as there are possible and impossible "objects." The "real" world is that of the range of existent objects, while, say, a "gold mountain" is a nonexistent but conceivably possible "object" in a possible world, and a "square circle" is a nonexistent and impossible "object."[11]

Russell (1973, 21–76) criticizes Meinong on the grounds that he violates the law of noncontradictions and thereby creates "category mistakes." It is impossible, Russell says, for anything to be both round and square, hence nothing can exist possessing both properties, and a square circle cannot be any sort of object. Russell argues further that if "gold" and "mountains" exist, then a "gold mountain" should be both "gold" and a "mountain" and it should exist. But there is no such thing, therefore a "gold mountain" does not and cannot reasonably exist.

However, Meinong would cryptically reply, a "gold mountain" is an object although it does not exist. Nonexistent "objects" are as "real" as existent objects; the only difference is that they do not exist. And, with respect to such impossible "objects," if they are contradictory there is no problem. That is why they are impossible, for if possible, then they would not be contradictory. Furthermore, to deny a "gold mountain" exists is automatically to assert the existence of *something*. Or concomitantly, to state that a fiction is not "real" is to assert that it has some sort of "reality" by the mere fact that there is an assertion *about* a fiction as *something*-to-be-asserted-*about*. There must have been some sort of prior "suspension of disbelief" in that fiction in order to consider it *as if* "real" before the possibility existed of denying it any ontological "reality."

Admittedly, Meinong continues, nonexistent "objects" are in a sense inferior to physical world objects, but this inferiority has nothing to do with their beingness or nonbeingness. It involves their incompleteness, because fictional or nonexistent "objects" invariably remain open and incomplete; that is, they will perpetually be incompletely described, for something more

can always be said of them. However, since the ultimate bounds of any and all possible "real worlds" from all possible perspectives are, like the bounds of all possible fictional worlds, indeterminately variable, knowledge of "real world" objects is placed at the same level as that of fictional "objects." So goes Meinong's strange world.

Intellection and Contemplation

Rescher and Brandom (1979, 33) stake the claim that reconstruction of Meinong's hypothesis "can be effected rather straightforwardly within the framework of a theory of nonstandard possible worlds." Assuming this premise, it follows that there can be indeterminate, self-inconsistent, and incomplete "objects" that nonetheless do not exist. I suggest that, in good Borgesian or Vaihingerian fashion, we consider these "objects" themselves as fictions. Thus, "square circles," "$\sqrt{-1}$," "Centaurs," and "gold mountains," though nonexistent, are mental "objects" that can be postulated for a Pythagorean, a quantum physicist, or any other fiction-maker. And all have, potentially, a chance of possessing correlates with whatever the invented "reality" is within the context of a given cultural framework.

In a comparable sense, and perhaps surprisingly, Tobias Dantzig, a mathematical "fictionalist" in his own right, speaks of the arithmetization of geometry initiatied by Descartes. He cites the example of the complex number

> which had its origin in a *symbol for a fiction,* [and] ended by becoming an indispensable tool for the formulation of mathematical ideas, a powerful instrument for the solution of intricate problems, a means for tracing kinships between remote mathematical disciplines.
> MORAL: FICTION IS A FORM IN SEARCH OF AN INTERPRETATION. (Dantzig 1930, 205)

The same can be said of all fictions: they are forms in search of interpretations, worlds crying out to be recognized. Hence the connection between Borges's and Vaihinger's "as if" hypostats, Rescher and Brandom's "non-standard semantically possible objects," and even Meinongian "objects," some or many of which, at one time or another, stand a chance of becoming "real" in the eyes of their beholder.

"Tlön, Uqbar, Orbis Tertius" (in which Borges refers to both Vaihinger and Meinong) provides a remarkable example of Dantzig's dictum. Following dinner, Borges (the narrator) engages in a casual conversation with his friend and collaborator,

Chapter One   Adolfo Bioy Casares. Bioy alludes to an article on the country of Uqbar in *The Anglo American Cyclopaedia,* and this article in turn refers to another planet called Tlön. After frustrated attempts to discover some literature on this curious region, Borges and his associate finally chance upon the eleventh volume of *A First Encyclopedia of Tlön,* which includes a description of the history, arts, metaphysics, and science of a totally imaginary world. The inhabitants of Tlön are absolute idealists in the Berkeleian sense. "Existence" begins with the postulate that the universe has no materiality and is nothing more than a projection of the subjective mind. Hence there are no nouns in the Tlönians' language, only verbs and adjectives. Since "reality" is wholly mental, there can be no legitimate science except psychology. Nor is there any consciousness of cause and effect, for everything created by the mind is mere association of ideas, in the sense of Hume. Furthermore, every mental state is irreducible. If a given mental state is named—classified—it is automatically falsified because all taxonomies are ephemeral and arbitrary. From this it might even be inferred that there are no "pseudosciences" in Tlön, but such an inference would be erroneous, for "pseudosciences" and metaphysical doctrines abound. That is to say, all hypothetical "realities" necessarily "exist" in simultaneity; they are conceived as a sum total of self-falsifying, *mutually cancellatory* affirmations and negations. It is judged, consequently, that metaphysics is a branch of fantastic literature; this must be so because the Tlönians have discovered that each system is no more than a reduction of all aspects of the universe to a handful of aspects. In this light, books should have a distinct function in Tlön, and they do. Since it is considered that all books are written by a single author, they are left unsigned. Novels contain a single plot with all its imaginable permutations. Philosophical works include both their thesis and their antithesis. A book "which does not contain its counter book is considered incomplete" (*L,* 13)

The only unthinkable metaphysical doctrine in Tlön is, understandably, materialism. "Reality" is considered by the Tlönians to be no more than temporal and sequential. Lacking spatial characteristics, "reality" therefore does not tolerate the self-identity of an object from one moment to the next. In Tlön there is only a myriad of successive differences, a self-identical single "Reality" being inconceivable. In other words, to paraphrase Goodman, Tlön is one "Reality," but it embraces a multiplicity of contrasting aspects; that is, Tlön consists of many "realities," and the collection of all of them is One. Consider the

following example of the absence of self-identity on Tlön: "Two persons look for a pencil; the first finds it and says nothing; the second finds a second pencil, no less real, but closer to his expectations" (*L,* 13). Mental preparedness—expectations—calls up the "reality" to be at a moment's notice.

Intellection and Contemplation

Finally, we are told that the idealism of Tlön is gradually overtaking our world. Its language is becoming our language, a fictitious past occupies what was our historical past. Mental "objects," in the Tlönians' conception of things, are endowed with the power to become what we would ordinarily consider "real."[12]

Perhaps, given all historical contexts, past, present, and future, Tlön is not a far cry from our own "reality." Borges implies so much. Referring once again to Meinong in "New Refutation of Time" (*OI,* 171–87), he states:

> Meinong, in his theory of apprehension, admits the apprehension of imaginary objects: the fourth dimension, say, or the sensible statue of Condillac, or the hypothetical animal of Lotze, or the square root of -1. If the reasons [for the refutation of time] I have indicated are valid, then matter, the ego, the external world, universal history, our lives also belong to that nebulous orb. (*OI,* 185)

Borges here speaks like a good fabulist and a hypothetical Meinongian. A universe in which a single writer writes all existing novels with all their possible permutations, in which each philosophical work must contain its own negation, in which all scientific hypotheses in their composite are self-refuting, is a universe viewed abstractly, ahistorically, and spatially—a view *sub specie aeternitatis*. The reasoning and logical nature in us strives to view things from such a perspective; in contrast, the sentient organism in us sees the world concretely, sequentially, temporally. The striking contradiction is that the Tlönians' mental world is exclusively temporal, while Borges describes Tlön as a simultaneity, from a totalizing vantage point "as if" he were outside looking in—and, of course, he is. This is a Vaihingerian fictive trick of the first order (as we shall note in the second section of Chapter Four, a comparable trick was employed by Einstein in constructing relativity theory).

Perhaps this Vaihingerian fictive trick is a sine qua non for doing science in the proper Western fashion. In this event, then, perhaps we can never know the world "out there," for we have

nothing to rely on besides fictive standards. Regarding this lingering problem, let us briefly consider Ernst Mach, physicist and philosopher of science, who lived about the same time as Vaihinger. Mach was a positivist, but unlike Vaihinger, he did not believe that atoms, for example, are convenient fictions. He believed categorically that they do not exist, for we cannot observe them directly. Mach's positivism was of the "phenomenalist" or Humean variety: all we can know about the world is what can be expressed as a result of our sensations. No philosopher had more influence on Einstein shortly after the turn of the century than Mach. But Einstein soon rejected Machian positivism, ironically at a time when a new crop of physicists, including Werner Heisenberg, adopted "phenomenalism" and became subjectivists. From that point onward, physics, and more specifically the Copenhagen interpretation of quantum theory, became increasingly subjectivist and subscribed to the belief that the observer necessarily interferes with the observed and hence cannot know objectively.

Machian phenomenalism entails a scientific quandary (as well as an everyday quandary for that matter). If sensations are the sum total of what we are, we still tend to nurture the hope of some guarantee that our sensations can be verified, for if not, there is no certainty that we can know anything at all. But if there is such a guarantee, then that guarantee, based on nothing more than sensations, requires that there be another guarantee that it be verified, and so on. As Hume demonstrated dramatically, induction can be no royal road to knowledge. Einstein's later contention with Mach's philosophy and inductivism stems from his Kantian idea that mathematical concepts transcend sensations and experience and are therefore effective instruments for scientific inquiry, though empirical evidence may not be immediately available.

In fact, Einstein—and here I extend his general world view—once commented on a lengthy passage from Russell's *An Inquiry into Meaning and Truth* (1940, 14–15). Russell suggests that we all begin with "naive realism" (the doctrine that things are what they seem). We believe that grass is green, stones are hard, and snow is cold. Physics eventually shows us that greenness, hardness, and coldness are not what our experience tells us, but something entirely different; they are only the effects of grass, stones, and snow on us. Thus "science seems to be at war with itself: when it most means to be objective, it finds itself plunged into subjectivity against its will" (Russell 1940, 283).

Russell finishes with the conclusion that naive realism gives way to physics, and physics, if true, demonstrates that naive realism is false. Hence "naive realism, if true is false; therefore it is false" (ibid.).

Einstein goes on to note that naive realism led to something akin to Mach's phenomenalism, which in turn gave rise to Hume's dilemma. But we have always garnered a desire for certainty; that is why Hume's message is such a crushing blow. The only alternative, it seems, is idealism. Of course, at the outset the thought of Berkeley and Hume appear contrary to scientific activity altogether. However, Einstein points out that Russell's passage "uncovers a connection." Kant believed he had the answer to Hume: if there is certain knowledge, then it cannot be empirical in origin, it must be grounded in reason; we have certain knowledge, therefore it is grounded in reason. However, Kant's "Gibraltars of Solid Reason," for example, geometry and the principle of causality, are known today to be much less determinate than Kant thought. Einstein continues, asserting that "the concepts which arise in our thought and in our linguistic expressions are all—when viewed logically—the free creations of thought which cannot inductively be gained from sense-experiences" (Einstein 1944, 281–87). Through no fault of his own, then, Hume created a danger, a "fear of metaphysics," which has come to be seen as a malady of contemporary empiricism. However, Einstein sees no threat from metaphysics; the scientist, he concludes, cannot get along without it.

F. S. C. Northrop (1949, 387–408) confirms Einstein's idealistic strain. Einstein believes that we can only grasp "reality" indirectly, by speculative means. Hence our view of "reality," Einstein readily admits, can never be complete, nor can it be final. It will forever be subject to change; sensory data will never cease to be reinterpreted, because theory determines what we see, not the other way around. Any and all attempts to explain "reality" inevitably culminate in changes in the mirror placed in front of that "reality." Joseph Agassi (1975, 119), philosopher of science, concludes, after discussion of, among others, Russell, Hume, and Einstein, that "science must remain at war with itself." That is to say, science must be perpetually in flux.

Thus we see that a scientist of the stature of Einstein, who more indefatigably than others persisted in his search for a unified theory that would finally bring order to the universe, and who maintained to the end that "God does not play dice,"

admits, nonetheless, to an inextricable tinge of idealism.[13] In fact, speaking of his space-time construct in relativity theory, Einstein (1934, 35–36) actually used the term "fictitious" to designate the nonempirical, hypothetical component of scientific theories. In this sense Einstein is inexorably drawn toward a conceptual framework not entirely unlike that of the Tlönians: our knowledge is mind-dependent before it is world-dependent.

At the same time, of course, Einstein never entirely relinquished his realist stand and embraced not naive realism but "scientific realism." He, unlike Vaihinger and nominalists, remained committed to the world; after all, he was a scientist. Borges, of course, has felt no such obligation, and hence his nominalist inclinations, which Einstein could never have embraced. Still, from the broadest possible perspective, parallels between Einstein's world-making and Borges remain. Borges repeatedly erects elaborate mental constructs, but he knows the fiction must soon end, that his relating a particular or a collection of particulars to the universal is ephemeral. So he demolishes his edifice, the fiction comes to a close, and we are left in our own perishable world. Einstein cannot engage in the fabulist's game of world-making. He knows his elaborate construct can never be complete. Nevertheless, he incessantly strives to perfect it. More importantly, with most contemporary physicists, he realizes what Kant and Newton did not: that nothing is permanent, not even the most elegant, exact, and faithful picture of the world. Like Borges's fictions, all world pictures will eventually be demolished. If every scientific theory except those accepted today is considered to a greater or lesser degree falsified, then there is virtually an infinitely greater probability that today's theories are false rather than true. I am sure Borges would be delighted with this one and only *Truth*.

In concert, the unexpected conjunction of Borges—as well as a host of other contemporary fiction-makers—with Vaihinger, Meinong, Goodman, and Einstein provides a stark, necessarily partial, sketch of a dizzying, labyrinthine universe of infinite complexity. In order to comprehend this universe by finite means, humankind, throughout history, has become inexorably incarcerated in dilemmas ("hyperfictions") of its own making, as we shall now observe.[14]

Chapter Two **A Predilection for Paradox**

This century could well go down in history as the "age of paradox." Beginning with the death of that great discloser of inconsistent premises and presuppositions, Nietzsche, and Georg Cantor's perplexing Chinese box of infinities, we have experienced Bertrand Russell's stopgap solution to paradoxes in set theory, the "limitative theorems" of Kurt Gödel, Alonzo Church, Alfred Tarski, and others, contradictory and non-Aristotelian logics such as those of Stephane Lupasco, Count Korzybski, and Rescher and Brandom. Then there is the strange case of quantum logic, in addition to the disconcerting effects of relativity theory and quantum uncertainty. On the other side of the ledger, paradox has surfaced in dadaism and other venerable "isms," the philosophy of the absurd, the notion of textual undecidability via Jacques Derrida, and in the art of, from among a growing number of representatives, Mavrits Escher, René Magritte, Marcel Duchamp, and Salvador Dalí. Finally, Borges must be counted among the chief fabulists of contradiction, which include the likes of Samuel Beckett, Franz Kafka, Eugene Ionesco, and Luigi Pirandello. In this chapter, I discuss paradoxes in general, and Borges's particular use of them, especially those of the Zeno variety, which will set the stage for future themes.

Chapter Two   *I don't like writers who are making sweeping statements all the time. Of course, you might argue that what I'm saying is a sweeping statement, no?*

—*Jorge Luis Borges*

According to the Eye of the Beholder

1 It has been said that paradox is truth standing on its head to attract attention, and that truth is paradox crying out to be transcended. The word comes from the Greek *para doxos,* meaning *beyond belief,* which is actually not befitting, for many paradoxes are the source of deep-seated convictions, if not "truth." More appropriately, then, we might say that paradoxes are trains of thought condensed into a point of time and space. Contemplating a paradox has been compared to meditating on a Zen Koan, gazing at a mandala, entering momentarily into the realm of the infinite. A world free of paradox is the stuff only dreams are made of, yet rationalism, even logic itself, in the final analysis "teaches us to expect some dreaminess in the world, and even contradictions" (Peirce 1960, 4:79). According to Kierkegaard, reason ultimately leads to paradox, and faith is needed to remedy it. But a paradox is not resolved by faith alone, nor can it logically be disposed of in many cases. It remains coiled at the very heart of our reasoning process.

Guillermo Sucre (1970, 469) correctly remarks that Borges's writing is a "fusion of contradictions." What he fails to note is that such fusion is inescapable, for the knowledge paradox inheres in all viable thinking, whether one be a nominalist, realist, idealist, Vaihingerian "fictionalist," Meinongian, or whatever. That is, the fusion of one's knowledge (what one believes one knows) with one's "meta-knowledge" (one's knowledge *about* one's knowledge) breeds paradox, for, like the "preface paradox" that marked the beginning of this book, at the second level one ultimately knows one does not, and cannot, be in possession of absolute "truth"; some of one's knowledge is, unfortunately, always either inconsistent or incomplete, yet at the first level one tends to persist in believing it may well be absolute. Rather than "truth" standing on its head, then, the knowledge paradox is "truth" as an asymmetric object looking at itself in a mirror. It sees its own inverse, its right side becomes its left side: its mirror image is its own falsity, and it can do no more than oscillate between the two poles of the contradiction ad infinitum.

A Predilection for Paradox

Richard Burgin (1968, 115) observes that Escher's work in visual terms compares favorably with some of Borges's cherished themes. Indeed, some of Borges's stories elicit what Douglas Hofstadter (1979, 94–95; 688–89) calls the "authorship triangle." There are three authors, A, B, and C. A exists in a novel by B, B in a novel by C, and, strangely, C is to be found in A's novel. This tangled triad is analogous, Hofstadter tells us, to Escher's well-known print of a hand drawing a second hand which is in turn drawing the first hand. How are such puzzles explained? Authors A, B, and C, necessarily unaware of their predicament, will, we must presume, happily tred through life believing they are real people. But another author, say, D, from a "meta-perspective," knows that they are mere imaginary beings. So assume D writes his own novel about those three unfortunate souls who think they are real people. Fine. But how is he to know that he is not a character in another novel by author E? We can gaze at Escher's hands drawing themselves and with confidence remark on the anomaly from our "superior" vantage point. We know that behind this print lurks Escher's invisible hand drawing the two appendages. But the problem is not resolved thus, for, like the knowledge paradox, from a "meta-perspective" there is still no guarantee that what we know is not false (or fictitious). There is no recourse but to oscillate between the two horns of the dilemma.

On speaking of such dilemmas, we are introduced to one of Borges's finest stories, "The Circular Ruins" (*L,* 45–50), which I will treat in some detail here in order that its paradoxical force may be effectively highlighted. In an exotic setting, a magician-priest arrives at the charred ruins of an ancient circular temple. The purpose which guided him "was not impossible, though it was supernatural" (*L,* 46). He wanted to dream a human being and insert him into reality. After failing in his initial attempt to dream a multitude of young boys and select from them the most promising candidate, he embarked on a second effort: to dream an individual, starting with the heart, and creating outward to the skeleton and finally to each of the innumerable hairs. After gradually accustoming this arduously dreamt boy to reality, the magician sent him downstream to the north "to be born." His son was now, for practical purposes, a part of "reality": in fact, "all creatures except Fire itself and the dreamer would believe him to be a man of flesh and blood" (*L,* 48). One night the magician was awakened by two boatmen who told him of another magician to the north who could walk on

fire without being burned. As any good father, the dreamer feared for the emotional well-being of his son, for if he meditated on his rare privilege and discovered that he was a mere image it would be humiliating. However, his thoughts were cut short, for a jungle blaze threatened from the south. The old man, cognizant of his imminent death, walked boldly into the "concentric" blaze only to realize "with relief, with humiliation, with terror," that the flames could not consume him, for "he too was a mere appearance, dreamt by another" (*L*, 50).

"The Circular Ruins" reveals one of Borges's strategies for creating contradictions, paradoxes, and infinite regresses, which will have a bearing on later sections of this inquiry. The first step entails a set of apparent oppositions. The magician came from the south, where he had dwelled in "one of the infinite villages upstream." On the other hand, after sufficiently preparing his "unreal" son for integration into "reality," the magician sent him downstream to the north, where there lay the ruins of "another propitious temple, whose gods were also burned and dead" (*L*, 47). The conditions of the son's environment are reciprocally identical to those of the magician. Only the infinitely repetitive trees of the jungle separate one temple from another. Hence, the spatial trajectories of father and son compose two symmetrical oppositions, up(stream)/down(south) and down(stream)/up(north), which structurally produce a "cancellation effect." As a result, the action of the story terminates simultaneously everywhere and nowhere: at the center of the charred ruins of a circular temple where the magician created his dream image. This symmetry of space reveals the fallacy of what Alfred North Whitehead (1925, ch. 4) calls "simple location." The story alludes not to precise geographic points but to vague notions of circular surfaces, which become almost as haptic as they are visual.

In contrast to these spatial indices, at the outset it appears that time is linear, and it accumulates with increasing torpidity. The magician required fourteen days to perfect the heart of his subject, one year to create the skeleton, a little less than two additional years to complete his project, and two more long years to prepare his son for "birth." The son's development, then, is first decelerated and finally halted altogether when the magician interpolates him into the world. However, this effort to annihilate the past is ultimately futile. Temporal recurrence is foretold by the magician's impression that "all this had happened before" (*L*, 45). If the obliteration of "simple location"

of space coupled with vague images of spatial circularity implies a denial of linear movement, concomitantly, the attempt to annihilate the past and establish an eternal "now" stems from an implicit attempt to deny temporal irreversibility.

The "invincible purpose" that drives the magician can be explicated on two levels: concrete and abstract. On a concrete level, the magician strives to coordinate his activities with those of his son. He daily prostrates himself at dawn and at twilight "before the stone figure, imagining perhaps that his unreal child was practicing the same rites, in other circular ruins" (L, 49). By means of these ritualistic acts, he gradually becomes "as all men," and his absent son is nurtured with the progressive diminution of his maker's own soul. When the magician's purpose in life is finally completed, he assumes that his son's immortality is now projected into the physical world, an event that at once symbolically represents the concretion of the "unreal" (dream) and the eternal coexistence of the "real" (physical) world. On an abstract level, the coexistence of "real" father with "unreal" son coheres with the symbolic coexistence of space and time. Spatial and temporal synchronicity portrayed in Borges's story is a condition quite unlike the linear temporal existence of the physical world. In this sense, physical existence, which presupposes human finitude, is opposed to the dream world of spaceless and timeless coexistence. In the material sphere of existence, the contradiction between life and death is presumably irreconcilable. On the other hand, in the nonmaterial order, governed by spatiotemporal synchronicity, this contradiction is nonexistent.

Consider the possibility that in "The Circular Ruins" a projection of spatio-temporal synchronicity into linear existence entails a symbolic abolition of the life/death dichotomy. This assumes an implicit attempt to overcome temporal existence wherein spatial hierarchy and temporal linearity predominate. In more concrete terms, the magician's "purpose" stems from a desire to make his "unreal" son part of tangible "reality" and vicariously to transcend mortality himself, for even though all fathers "are interested in the children they have procreated," this interest is at the same time self-interest. Therefore, the constraint in the text subjected to potential restructuration is the life/death duality, perhaps the most intransigent of all. For obvious reasons, the protagonist is a "magician" and the story reads like a "myth."

The relation between father and son ("reality" and dream) can be illustrated by an abstract scheme (figure 1), in which

Chapter Two the sequential and parallel planes intersect where there is potential movement in the narrative toward a more complex level of organization. The desired goal entails actualization of relations of similitude between father-son and "reality"-dream. By inserting dream image into "reality," the son can become a "man" and the magician can vicariously transcend the finitude of his physical existence. In order to accomplish this goal, the magician must activate a fusion of opposites wherein the son's *timelessness* predominates over the father's *temporality* and the father's *essence* over the son's *materialessness*. However, the "logical" end prevails: the magician realizes he is an integral part of dream existence, which discounts the son's supposed entry into "reality," and the "unreal" enjoys synonymity with the "real."

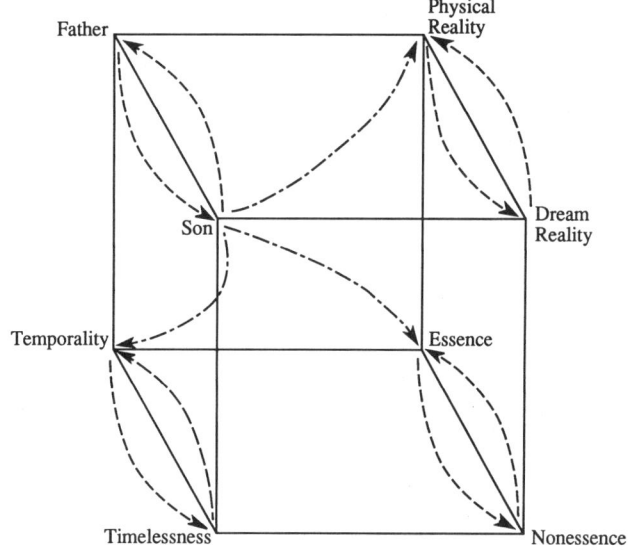

(1) Horizontal and vertical lines represent sequential relations.
(2) Diagonal lines represent (negative) parallel relations.
(3) ._._._ is the desired goal.
(4) _____ is the "logical" end.

Figure 1

According to the reading I have proposed, this impossible intersection of the "unreal" and the "real" becomes manifest at the end of the story. The magician assumes that his monumental task is completed, but when the parallel and sequential axes converge, the paradox underlying his project potentially be-

comes apparent. His status as the object of yet another dream is obviously a proposition embedded in his mind, since his own maker had instilled in him, as he in his son, complete oblivion of his apprenticeship. Hence, from the very beginning, it may be conjectured, the magician's grand design is doomed to failure. In the first place, he strives to force the dreamt image into his own supposedly tangible form of existence in order to concretize the sequential chain of mental events (dream "reality") that are the product of unlimited semiotic activity and to establish lines of similarity where ordinarily there would exist only lines of opposition. In other words, he tries to make dream "reality" denote something other than what it would ordinarily denote. In the second place, realization of the magician's desire would be equivalent to a desiring subject's becoming part of the imagined world he has created and at the same time a prisoner of/in his own desires. The fusion cannot be actualized, however, and what the magician presumed to accomplish at one level backfires at another level.

A Predilection for Paradox

To determine more precisely the nature of the magician's dilemma, let us return to the implicit purpose guiding his action. After the magician's preliminary effort to create a "real" son fails, he realizes that his project will be much more arduous than "weaving a rope of sand or coining the faceless wind" (*L,* 47). This passage reveals two metaphorical (oxymoronic) images, which on a local level represent the impossible conjunction of distinct classes of things: rope (fibered) out of sand (nonfibered) or coin (malleable-solid) out of wind (nonmalleable-nonsolid). The magician now attempts to construct a solitary dreamt image by means of another approach. The problem is that to integrate the attributes of this image into his own world logically implies a simultaneous rupture of boundaries. In other words, two distinct classes, *A* and *B,* are governed by different logical orders, and they cannot be integrated while maintaining intact the logical order of either *A* or *B,* but both, on becoming members of the same class, must be subjected to a different order. Hence, the magician cannot conjoin two distinct spheres of existence into his own without altering both. If, on the other hand, the magician had conceived of his dream world as does primitive man, as merely another facet of the same "reality," his project would nonetheless have been equally futile. For to make a dream coexist with "reality" would be nonsensical, given the fact that in the primitive's animistic conception of things, the two entities could not represent an intractable dualism in the first place.

Fire might have been construed as a potential mediator between the "reality" of the magician and the "nonreality" of the boy, since fire symbolically "converts" essence to nonessence (matter to energy). Following this metaphorical line of reasoning, the magician would be attempting to reverse the process and convert his "unreal" son (nonessence) to "reality" (essence). Moreover, only fire would be able to discern the created being's lack of essence, since it cannot consume that which is the final product of its consummatory process. In fact, fire appears as an earthly god in one of the magician's dreams and offers to give life to his inert dream image. However, the fire deity is helpless against that over which it presumably exercises dominion: its very sanctuary, as in centuries past, is destroyed by fire. This destruction recapitulates the paradox inherent in the magician's project. That is, the god of fire is the "symbolic," or "archetypal," expression of fire, and as such it rests at a distinct level of organization. The symbol can be representative of fire but cannot coexist on the same logical level as fire; it cannot be fire itself. When the magician assumes he possesses the ability to annihilate the boundaries between logical categories, all distinctions between symbol and referent, dreamer and dreamt, become nonexistent, and he loses his capacity, as *Homo symbolicus,* to create an ideal world that rests in contradiction to "real" reference.

"The Circular Ruins" is merely one exemplary instantiation of Borgesian paradoxes. Like the Aleph itself, most of these paradoxes entail an untenable collaboration of the infinite and the finite, time and timelessness, continuity and discontinuity, the One and the Many. These are "hyperfictions" in the extreme; that is, they imply oxymoronic collusion at the deepest level. Weaving a rope of sand or coining the faceless wind serve as microcosmic embellishments of the larger tragedy being played out, which is equally oxymoronic, but, in addition to its poetic qualities, it is also metaphysical—cosmic metaphors, which place Borges's fictions in the same orb with the range of Vaihingerian fictions and scientific models-fictions.

Regarding further the nature of Borges's paradoxes, Bruce Lorich (1973, 53)—and he is certainly not alone here—disparagingly proclaims that most of Borges's fictions are "heretical texts for solipsists." The Argentine writer, he declares, "has created a complex, labyrinthine puzzle of paradoxes in which pedantic psyche-twisters, related only to themselves and to a philosophical stance, appear more regularly than do ingenious

truly fictional artifices. Most of his so-called fictions are intellectual games."[1] Lorich, I believe, is correct on one point: Borges's fictions are intellectual games. But they are far from trivial. The fact is that all paradoxes potentially bear on "truth," but in a negative manner, for negation (i.e., Nietzsche's "lie") is one of the prerequisites for their existence. For example, a variation of the Liar paradox, "This sentence is false," entails negation, since it says of itself that it is *not* true. It also entails two other conditions necessary for paradox: self-reference and infinite regress. The sentence self-referentially speaks of its own lack of truthfulness, and hence if it is true, then it is false, and if so, then it is true, and so on, ad infinitum.

These three conditions—negation, self-reference, and infinite regress—should reveal many pseudoparadoxes being passed off as logical paradoxes in Borges criticism and elsewhere. As a case in point, the injunction

> IGNORE THIS SIGN

places the reader in a quandry, but it is not logically paradoxical, since, rather than reflect upon itself, it includes the reader; it constitutes merely a "pragmatic" paradox. That is to say, the reader, having already read the sign, cannot ignore it so as to obey the injunction, hence there is no infinite regress. Nor can self-reference alone create paradox, in spite of the overenthusiastic claims of certain contemporary literary critics of poststructuralist ilk. It might simply create an absurd situation comparable to Lewis Carroll's mythical island whose inhabitants earn a living by taking in each other's washing. With respect to Borges's work, for example, critics have often commented on the "paradox" inherent in *History of Eternity,* since eternity is atemporal and hence incompatible with the notion of historicity. However contradictory this title may be, it is not paradoxical, for there is no infinite regress.

The essence of Borges's genius lies in his recognition of false dichotomies and his uncanny ability to dramatize paradox within an ethereal, timeless framework. Borges seems to suggest that if the universe operates consistently at all, it operates consistently on paradox, and insofar as it does so, the universe may be rational instead of absurd, a rationalism, nonetheless,

Chapter Two

edified upon inconsistent premises. Paradox, like infinity, is contingent upon simultaneity. "This sentence is false" must, logically speaking, be construed at once as both true and false. However, the mnemonic function of the brain is linear. It entails sequential expression and analysis, which contradicts the simultaneity essential for Borges's and other paradoxes to emerge. Properly perceiving paradox, then, involves the perceiver's oscillating between the *either* and the *or,* an oscillation in time, but, since it goes nowhere so to speak, it is timeless. It has no tenses of past and future, for the oscillations in their composite are static.

Paradox viewed from two perspectives, the temporal and the timeless, can be illustrated by what is called the "prisoner paradox." It is Sunday. A group of prisoners is told by the warden that they will be executed on one day of that week, but they will not be informed which day it will be until the arrival of that very day, hence it will be a surprise. These prisoners happen to have found a quite astute lawyer. He reasons, after some deliberation, that assuming the warden to have told them the truth, they cannot be executed, for if the fatal day is to be Saturday, then it cannot be a surprise, since it will be the only day remaining. By this mode of reasoning neither can it be Friday, for Saturday now having been eliminated, Friday is no longer a candidate. The same can be said of Thursday, and so on down to Sunday. Therefore they cannot legitimately be executed. Now, there seems to be a flaw somewhere, and there is, but it has nothing to do with logic. The problem is that the lawyer's reasoning is strictly by logical means; he can certainly afford to be logical, for his life is not at stake. In contrast, the prisoners' very existence is in jeopardy. They are rightly concerned over how much time remains of their life, and time is precisely the issue here. The lawyer's logic is timeless, and within this framework the paradox springs forth in full force. But the prisoners, their emotions having understandably taken precedence over their reasoning faculties, exist in time, hence try as their lawyer may to convince them otherwise, they cannot reason away their expectations of an unexpected moment announcing their death. Trapped inside their particular mind-set, they are condemned to temporality.

Existentially the human organism cannot help but project into the future and carry within itself conglomerate memories of the past, for without such memories, there is no future anticipation. Such would be a rudimentary form of animal conscious-

ness, which, as Borges often reiterates in his works, is situated outside time. We humans, in contrast, are caught within a temporality that, for better or for worse, generates quandaries necessary to our very existence. For example, the narrator's obsession with the twenty-centavo coin in "The Zahir" exhibits our unwillingness to let go of the past: he simply cannot forget the coin, yet at times he feels so confident he can forget it that he "deliberately recalled it to mind" (*L*, 160–61). Here Borges reveals the subtle "forgetfulness paradox." The narrator gives himself the injunction "Don't think about the Zahir" or "Forget the Zahir." But if he does not think about it, then he runs the risk of forgetting what it is he is supposed to forget and he might think about it; therefore he must maintain mindfulness of it so as not to forget he must forget it. This pragmatic paradox depends as much on memory as does the "prisoner paradox" on anticipation. And both are existential rather than purely logical, time-bound rather than atemporal.

Interestingly, in Borges's "The Secret Miracle" (*L*, 88–94) we have another variation on the prisoner paradox. Jaromir Hladik is condemned to die before a firing squad at dawn, the twenty-ninth of March, 1943. He "infinitely anticipated the process of his dying, from the sleepless dawn to the mysterious volley" (*L*, 89). He imagined himself dying hundreds of deaths "in courtyards whose forms and angles strained geometrical probabilities, machine-gunned by variable soldiers in changing numbers, who at times killed him from a distance, at others from close by" (*L*, 89). One day he reflects that

> reality does not usually coincide with our anticipation of it; with a logic of his own he inferred that to foresee a circumstantial detail is to prevent its happening. Trusting in this weak magic, he invented, *so that they would not happen,* the most gruesome details. Finally, as was natural, he came to fear that they were prophetic. Miserable in the night, he endeavored to find some way to hold fast to the fleeting substance of time. He knew that it was rushing headlong toward the dawn of the twenty ninth. He remembered aloud: "I am now in the night of the twenty second; while this night lasts (and for six nights more), I am invulnerable, immortal." (*L*, 89–90)

Hladik is a writer who has never been satisfied with the fruits of his craft. Perhaps his most successful book was the one entitled, significantly, *Vindication of Eternity.* The first volume "gave a history of man's various concepts of eternity, from the

immutable being of Parmenides to the modifiable past of Hinton" (*L,* 90).² The second denied that "all the events of the universe make up a temporal series, arguing that the number of man's possible experiences is not infinite, and that a single 'repetition' suffices to prove that time is a fallacy. . . . Unfortunately, the arguments that demonstrate this fallacy are equally fallacious" (ibid.).³ Indeed, with such remarkable precedents, how could Hladik possibly be overtaken by time?

To "redeem himself" of his past failures, Hladik has been writing a play in verse entitled *The Enemies,* which remains unfinished. He petitions God that he be granted an extra year to finish it before his execution. It is granted. Then, on the twenty-ninth, before the firing squad and at the moment the command is given to execute him, the "physical universe comes to a halt": Hladik's "weak magic" had triumphed. For the duration of one year, motionlessly, secretly, he then "wrought in time his lofty, invisible labyrinth. . . . nothing hurried him. He omitted, he condensed, he amplified. . . . He concluded his drama" (*L,* 94). Then he was shot at dawn, on the twenty-ninth.

For the prisoners, existence took precedence over logic. For Hladik, like the prisoners' lawyer, logic took precedence over existence. The lawyer's reasoning was helpless in the face of "reality," since the prisoners were sure to die on the designated day. Hladik's reasoning powers, in contrast, *became* "reality" in much the sense of scientific theories, which, purely imaginary mental constructs in the beginning, can subsequently become in the minds of their believers synonymous with the "real" (i.e., the Newtonian-Cartesian machine model of the universe and other such models and metaphors). We shall now examine Zeno's incisive proofs and their significance for Borges, which will ultimately have a bearing on contemporary physics, to be discussed in Chapters Five and Six.

It is in changing that things find repose.
—*Heraclitus*

To Reach the Unreachable

2 Zeno of Elea, a disciple of Parmenides, confounded his contemporaries with four apparently irrefutable arguments. Three of them, known as the paradoxes of *dichotomy,* of *Achilles and the Tortoise,* and of the *arrow,* are most relevant to Borges's work.⁴ According to the first, if we desire to walk one

mile, we must first walk half the distance, one half of a mile, and then half of that, one quarter of a mile, then one eighth of a mile, one sixteenth of a mile, and so on, ad infinitum.[5] There is an interesting lesser-known alteration of this paradox that increases rather than decreases the magnitude of the numbers. If after walking one half of a mile we count "one," then after the successive one quarter of a mile we count "two," then "three," "four," and so on, we must eventually count an infinity of numbers.[6] According to Zeno's second paradox, in Borges's own terms (*OI*, 109–10)

> Achilles runs ten times faster than the tortoise and gives him a start of ten meters. Achilles runs those ten meters, the tortoise runs one; Achilles runs that meter, the tortoise runs a decimeter; Achilles runs that decimeter, the tortoise runs a centimeter; Achilles runs that centimeter, the tortoise, a millimeter; Achilles the Nimble-Footed, the millimeter, the tortoise a tenth of a millimeter, and so on ad infinitum, with Achilles never overtaking the tortoise.

Borges (*OI*, 110–11) attributes to Aristotle a variation on this theme. Aristotle asks us to assume that the class of men and an individual man have common attributes; they have, if nothing else, the attribute of humanity. Taking the combination of these two, it will be necessary to accept a third, then a fourth, and so on. Borges explains it thus (*OI*, 111):

> Let us postulate two individuals, a and b, who make up class c. Then we shall have
> $$a + b = c$$
> But also, according to Aristotle,
> $$a + b + c = d$$
> $$a + b + c + d = e$$
> $$a + b + c + d + e = f \ldots$$
> Two individuals are not actually needed; the individual and the class are enough to determine the third man postulated by Aristotle. Zeno of Elea uses infinite regression to deny movement and number; his refuter uses infinite regression to deny universal forms.

Borges, it bears mentioning, provides additional variations on this paradox that need no further comment here, for the point is now obvious. Concerning the final of the three

A Predilection for Paradox

paradoxes, an arrow in flight at any given instant is at rest, as if with a row of high-speed cameras we were to take a series of shots of it. At each of these instants the arrow occupies its own space and no more, so it is not moving but stationary.[7] But what is true of the arrow at each moment is true of it throughout the entire period; hence during its flight it is never moving but stationary. In a strange manner, "The Secret Miracle" is relevant to this paradox, for assuming Hladik's mind, and hence his reality, to become purely logical machines at a given instant, the bullets in the rifles aimed at him, everything in his world, would be entirely static. Time would cease to flow.

Russell (1926, 183) remarks that "Zeno's arguments, in some form, have afforded grounds for almost all the theories of space and time and infinity which have been constructed from his day to our own." The theories of space, time, and infinity Russell alludes to are, in light of the discussion in Chapter 1, "mind realities"—the most radical of which are "hyperfictions." In the spirit of Russell, Borges, commenting on and discarding various attempts to dissolve Zeno's paradoxes, especially that of Achilles and the tortoise, juxtaposes the world of idealism ("mind-reality") with the empirical world ("sensory-reality"). He then proposes that "Zeno is incontestable unless we confess to the ideality of space and time. Let us accept idealism, let us accept the increasing concreteness of what is perceived, and we will elude the multiplication of paradoxes into the abyss" (*D*, 120).

This is perplexing. The chasm between "mind-reality" and "sensory-reality," given Eddington's "schizophrenic" desk, is apparently unbreachable, but Borges seems to be telling us that we can and should embrace both, as alternative worlds. Perhaps this is inevitable, given our Eleatic inheritance. The trend toward such a widening of the abyss between the two "realities" actually began with the Greeks, for whom, Russell observes, geometry was practically the only known science. There was no experimentation, no attempt at empirical verification or falsification. The Greeks constructed a purely logical world, thus liberating the imagination and denying the world as it was ordinarily perceived. Even upon properly considering the nature of the Greek's mind-dependent formulations, it still appears absurd that Zeno, through his "mind-reality," might actually have believed motion to be impossible. Granted, his skeptical conclusions follow logically from an ingenious dialectical argument, but to consider them to be empirical fact borders on madness.

A Predilection for Paradox

Yet, "it is not difficult to suppose that he really believed his arguments to be irrefutable" (Benardete 1964, 2-3).

Over the long haul, then, the Eleatic arguments are, properly considered, pure acts of mind. In comparable fashion, modern mathematics, the language of the physical sciences—and therefore, some still presume, of "reality"—is a vast set of mind acts. In its very origin, mathematics is not empirical, i.e., a matter of inferences from the initial act of counting actual objects in the world, as J. S. Mill and others have asserted. So it need enjoy no necessary correlation with "sensory-reality." Dantzig (1930, 123) points out, in this light, that the problems revealed by Zeno and the Eleatics "are not of the type to alarm the pure mathematician—they do not disclose any logical contradiction, but only sheer ambiguities by admitting that the symbolic world in which he creates is not identical with the world of his senses." The act of mind is a free creation, as Einstein asserts, or a fiction (in the sense of Vaihinger and Borges). As such, a given mental construct may, perhaps by way of luck more than well-reasoned judgment, correspond to "sensory-reality." And we smugly proclaim that the abyss is closed. But our security is ill-founded, if we lend an ear to philosophers of science like Paul Feyerabend, Thomas Kuhn, Norwood Hanson, and Michael Polanyi, for whom the world we "see" is not the world of raw sense-data.[8] On the contrary. Much in the sense of Goodman's world-making, we are to a large degree programmed to "see" the world "as" such-and-such, for our "seeing" entails a making, or remaking as it were, more than a finding.

With respect to mind-acts, infinity, and Zeno's paradoxes, at the beginning of "Avatars of the Tortoise" (*OI*, 109–15), we are introduced to Nicholas de Cusa, who in *Of Learned Ignorance* conceived of "a polygon with an infinite number of angles in the circumference and wrote that an infinite line would be a straight line, would be a triangle, would be a circle and a sphere" (*OI*, 109), all of this being possible if the curvature and the number of lines were infinitesimal and the line infinitely long. This, of course, is the method of analytic geometry in good Vaihingerian fashion. The mind-dependent curve is conceived *as if* it consisted of an infinite number of infinitesimally small straight increments joined together, and for all intents and purposes the curve is computed *as if* it were an empirical entity. A geometer could be more true to form by "representing" the "curve" as a series of visible straight increments, but this would

not satisfy the practical man, the engineer, who must apply his calculations (mind acts imposed on the world). We usually prefer the "appearance" of "reality," generally forgetting that a fiction was used to actualize that very appearance, and, while we might not remain blissfully ignorant of the paradoxes underlying it, we pay these paradoxes little or no mind when engaged in our practical daily affairs. In other words, we have for so long been trained in using these and other such fictions that we have come to prefer the substitute to the genuine article.

Borges repeatedly, and at times painfully, strives to call our attention to what lies behind our so-called rational thought. It can be said that he uses Zeno's dialectics, but for a distinct purpose. The latter wanted to prove logically that Parmenides' idea of movement being impossible is true, and he did so by a purely mental act. Borges does the same, while offering a historical view of the paradox, and at the same time he strives to demonstrate that the world cannot be adequately accounted for solely by the use of logic and reason, for the ultimate consequences of such use inevitably leads to absurdities (Zalazar 1976, 102–3). The point is well taken.

Entering now into the technicalities of Eleatic philosophy regarding Borges's prose, Zeno's paradoxes, it must be said, depend upon a linear progression. Such progression is found in a mental game created by Borges in "Death and the Compass" (*L*, 76–87), which culminates in a one-dimensional Zenoesque conceptualization as an alternative to more complex labyrinthine paradoxes. In this story, Inspector Lönnrot, the super-calculator, infers from his reading of Jewish texts, from number magic, and from geometry that after three enigmatic homicides a fourth murder is inevitable. He determines its exact location on a map, at the Villa of Triste-le-Roy, and appears there at the precise time he calculated that the murder is to be enacted. The villa

> abounded in pointless symmetries and in maniacal repetitions: to one Diana in a murky niche corresponded a second Diana in another niche; one balcony was reflected in another balcony; double stairways led to double balustrades.... through anterooms and galleries he [Lönnrot] passed to duplicate patios, and time after time to the same patio. He ascended the dusty stairs to circular ante-chambers; he was multiplied infinitely in opposing mirrors; ... *The house is not large,* he thought. *Other things are making it seem larger: the dim light, the symmetry, the mirrors, so many years, my unfamiliarity, the loneliness.* (*L*, 83–84)

Scharlach, the murderer, appears with two accessories who disarm and handcuff Lönnrot—the intended victim. Then there is a brief exchange of words between Lönnrot and Scharlach, during which the latter explains how Lönnrot's ratiocination was his own undoing. His purely formal construct that he believed could solve the murders did not correspond to his perceived world but to another artificial world, created by another mind, Scharlach's. Both worlds are symmetrical, like the labyrinthine villa. They mirror each other, as do the very names of the antagonists, Lönnrot and Scharlach. At the termination of this exchange, Lönnrot avoids Scharlach's eyes, as if to negate the symmetry evinced by the presence of the two men, and finally, he proposes an alternative to Scharlach's labyrinth. "In your labyrinth there are three too many lines," he begins.

A Predilection for Paradox

> "I know of one Greek labyrinth which is a single straight line. Along that line so many philosophers have lost themselves that a mere detective might well do so, too. Scharlach, when in some other incarnation you hunt me, pretend to commit (or do commit) a crime at A, then a second crime at B, eight kilometers from A, then a third crime at C, four kilometers from A and B, half-way between the two. Wait for me afterwards at D, two kilometers from A and C, again halfway between both. Kill me at D, as you are now going to kill me at Triste-le-Roy." (*L*, 86–87)

Lönnrot, of course, errs once again. Scharlach's labyrinth, with its multiple staircases, etc., is actually three-dimensional rather than constructed, as is the conventional maze, along a two-dimensional plane. This labyrinth is confusing, alienating, and maniacal; it is analogous to the indescribable, incomprehensible world of our sensory experience in all its complexity. In contrast, Lönnrot's alternative, a linear paradox, is of the most elegant simplicity, but it is mere fiction, a mental world. In spite of his vague allusion to Zeno, Lönnrot could no more prevent the bullet from reaching him in such a labyrinth than could the sentient Hladik—without the grace of God, we must suppose— stop time for a year. Yet Lönnrot is a helpless and hopeless reasoner, destined to perish in a mind-dependent world of his own making. On the other hand, Scharlach's response reveals his confidence in his own mental game: "The next time I kill you, . . . I promise you that labyrinth, consisting of a single line which is invisible and unceasing" (*L*, 87).

A geometer's line drawn on a graph is *in reality* not invisible; if it were, he could not carry on with his work; its invisibility

is a temporary Vaihingerian fabrication. And the line is unceasing only in an axiomatic sense, for the eye, limited to its sensory world, is incapable of determining the line's continuing ad infinitum. The line, *in reality*, is useful for geometrical exercises, but it is not the reality of the geometer's graph; his reality is abstract, purely mind-dependent, yet it is capable of patterning and governing *the real*. In contrast, Lönnrot, it appears, is destined perpetually to commit what Alfred North Whitehead (1925, ch. 4) terms the "fallacy of misplaced concreteness." He desires to interject an abstraction into his experienced world *as if* it were *really real*.

It is now propitious to introduce provisionally two related topics, the details of which will be fleshed out in Chapter Six: (1) the relationship between Zeno (or Borges) to quantum theory, and (2) the problem of infinity. Zeno's argument is in a sense challenged by Werner Heisenberg's uncertainty principle, according to which either the position or the momentum of a particle may be known, but not both in simultaneity. Restated in terms similar to the arrow paradox, if a particle is moving, it cannot be assigned a definite spatial coordinate, and if it is assigned such a coordinate, it cannot at that instant be in motion. This in essence also reveals Bergson's (1964, 328) refutation of Zeno: "all movement . . . is either an indivisible bound (which may occupy, nevertheless, a very long duration) or a series of indivisible bounds." In other words, all movement is either considered as a dense continuum (of becoming) or as a discrete series (such as the natural integers, successive straight lines forming a curve, or instants of time). It appears that in a sense we can have our cake and eat it, too. The notion of an object moving in a dense continuum gives Heraclitus the upper hand, but its motion, seen as a discrete series of fits and jerks, compels us to lean toward Parmenides and Zeno. Each pole of the uncertainty principle affords us a distinct picture; both poles in their own right are correct.

But this would be an oversimplification, for the crux of the issue involves the *type* of infinity we are speaking of. Infinity comes basically in two shapes: *actual* infinity and *potential* infinity. We tend to experience a chalk mark on a blackboard as continuous (an actual infinity) but it is not, for it merely fails to reveal noticeable gaps to the naked eye. It is no more than the visible expression of a potential. Zeno's paradoxes are predicated on the concept of a potential infinity (a never-ending succession of steps). However, since the implication of actual

infinity underlies this concept, Zeno is not a finitist, and finitistic arguments, which merely banish the idea of the actual infinite altogether, cannot legitimately dispel him (Benardete 1964, 13–20).

Does Borges allude in his stories to a potential or actual infinity? Waldo Ross once asked Borges if it was not possible to apply his idea of one's interior world, as found in "The Library of Babel," "The Aleph," "The Zahir," and other stories, to the concept of an actual infinity. Borges responded: "That question is too metaphysical. I am not a metaphysicist. But in any case I can tell you that the series of mirrors [or the mirror in The Library which duplicates all appearances, for example] is infinite. Aristotelian logic would be impossible here. The series of mirrors loses itself in the infinite" (Ross 1975, 280). Ross continues, conjecturing that if we contemplate Borges's image sufficiently, we shall conclude that it is neither static nor dynamic, or inversely, it could be simultaneously both static and dynamic. Ross's comment is essentially that of quantum uncertainty. The subatomic "event" is neither particlelike nor wavelike, or it can be looked upon as both particlelike and wavelike. In addition, the mirror in Borges's vast Library leads many to infer that the Library is not infinite, for if it were, then why this illusory duplication? The narrator, on the other hand, prefers to dream that the polished surfaces of the mirror "represent and promise the infinite" (*L,* 51). This being the case, Borges is speaking of the potential infinite. In contrast, the narrator later tells us that when he dies he will be thrown over the railing and "dissolve in the wind generated by the fall, which is infinite" (*L,* 52). There seems to be no doubt that Borges is here speaking of the actual infinite. Moreover, if the Library consists of a set of discrete galleries, then the question arises as to whether or not there can be an actual countable infinity of them. The infinitist says yes; the finitist says no. The argument, perhaps, will continue ad infinitum.[9]

From Borges's view, we are asked, at the end of "Avatars of the Tortoise," to admit with the idealists that the nature of the world is hallucinatory, but he also asks us to do what they have not done: search for the unrealities that confirm that nature. He declares that we shall find them in Zeno and the antinomies of Kant. And, he concludes, "We (the undivided divinities that operate within us) have dreamed the world. We have dreamed it strong, mysterious, visible, ubiquitous in space and secure in time; but we have allowed tenuous, eternal interstices

of injustice in its structure so we may know that it is false" (*OI,* 115). This interpretation of idealism apparently admits of a continuum where everything is possible. However, the interjection of paradox, forcing us to realize that this hallucinatory world is artificial, contradictorily confuses the discrete with the continuous. Interestingly, while Zeno needs the continuum to support his notion that a discrete series can never end, Borges needs the concept of discontinuity to remind him that the ideal world is repeatedly falsified.

A comparable topic found throughout Borges's prose is another type of series analogous to a set of Chinese boxes (to be discussed in another context in the fifth section of Chapter Six). For example, commenting on the dreamer in "The Circular Ruins," Borges (*A,* 267) says: "In my opinion, a speech implies a speaker and a dream, a dreamer; this, of course, leads to the concept of an endless series of speakers and dreamers, an infinite regress, and may be what lies at the root of my story." In other words, the magician's "real world" of experience entails continuous time and space, but his dream world is a series of static flashes in which space and time are quantized, and Zeno's arrow paradox seems to apply. However, just as the magician thought he had successfully interpolated his dreamt son into the "real world," so his dream, in memory, becomes an extension of the continuum, as has presumably the dream of the dreamer dreamt by another. The problem is not solved; it merely dissolves when one horn of the dilemma is lopped off.

The player in "Chess" (*OP,* 124–25), a poem, moves his pieces at will, but he is moved by God, who is in turn moved by a higher God, ad infinitum. This example also images the continuous and the discrete. The world of chess is in essence a set of squares within an all-encompassing square populated by discrete pieces scattered throughout, and each move of one piece is a "quantum jump." The game, however, is rendered continuous by the master player. Her focal attention rests on the entire board as an analog whole, a *Gestalt,* as it were, each discrete item having been pushed to the periphery of her total awareness. And her own puppeteer perceives her in the same fashion (somewhat like Scharlach observing Lönnrot).[10] "The Sect of the Phoenix" (*L,* 101–4) tells of a secret society that passes on a rite and a set of unutterables from generation to generation until both have supposedly become instinctive: the many become one and the one becomes as an unconscious dream. That is, submersion into consciousness of a sacred rite implies the

discontinuous (subsidiary awareness of events) becoming continuous (tacit knowledge of events as a *Gestalt*).[11] And finally, "The Approach to al-Mu'tasim" (*A,* 45–52) presents the description of an imaginary novel in which a student becomes obsessed with his search for al-Mu'tasim, who is God. At the same time, it is implied that "the Almighty is also in search of Someone, and that Someone of Someone above him (or someone simply indispensable and equal), and so on to the End (or rather, Endlessness) of Time" (*A,* 50).[12] This elongated set of (discrete) Chinese boxes representing searchers and searched doubles back on itself such that, paradoxically, the beginning is the end and the end is the beginning: the (continuous) circle is closed. To paraphrase Borges, the collection of all searchers is one searcher.

That Borges's very concept of dream is strangely Zenoesque is nowhere more evident than in "The Dream of Coleridge" (*OI,* 14–17). The English writer, in 1797, dreamed the lyrical fragment "Kubla Khan" but did not publish an account of his dream until 1816, as a justification for the unfinished poem. Twenty years later a Persian book, *General History of the World* by Rashid al-Din, which dates from the fourteenth century, when it appeared in Paris, confirms that Kubla Khan once built a palace according to a plan he had seen in a dream. A thirteenth-century Mongolian emperor dreams a palace and builds it; an eighteenth-century poet—who could not have known of the dream—dreams a poem about the palace! The first added a palace to "reality"; the second a poem. The similarity of the symmetrical dreams and the enormous length of time involved, says Borges, "reveals a Superhuman performer" (*OI,* 17). Outlandish as this might appear, Borges goes on to suggest that the purpose of this long-lived being is not yet finished, for someday "some reader of 'Kubla Khan' will dream, on a night centuries removed from us, of marble or of music. This man will not know that two others also dreamed. Perhaps the series of dreams has an end, or perhaps the last one who dreams will have the key" (ibid.). Borges surely does not believe his own words, we must surmise. And it appears that he does not, or so he indicates, on offering an alternative explanation hardly less credible than the first: "Perhaps an archetype not yet revealed to man, an eternal object (to use Whitehead's term), is gradually entering the world; its first manifestation was the palace; its second the poem. Whoever compared them would have seen that they are essentially the same" (ibid.).

Chapter Two

Infinity is a possibility, since the series of dreams might well have no terminus. The series is discrete, however: each dream is an increment, with an indefinite lapse of time in between. So, like the arrow paradox, there cannot logically be any transition from one dream state to another. If the arrow can leave the bow in the first place—which Zeno denies—it will never strike the target. Yet it does strike it, for the Buddha at least, who enjoys a timeless vision: he sees the arrow leave the bow, trace an arc through the air, and hit the target dead-center, all in a flash. Borges's "Superhuman performer," like the Buddha, must possess the same vision: all dreams are one dream, all men are one man; the world, ultimately, is One—the distinctive Heraclitus, in this conception of things, is therefore an illusion.

Chapter Three **The Demise of Totalizing Quests**

Viewed through Borges, all logical, mathematical, and metaphysical doctrines seem to culminate in paradoxes of one sort or another. Actually, such cosmic knots have been with us at least since Heraclitus and Parmenides began taking potshots at each other. The original question was essentially, Is static Being mere illusion or an invariant Divine Principle? Some early philosophers, endeavoring to overcome this perplexing conflict, surmised that Being is manifested through immutable substances. Atoms, the smallest units of matter according to Leucippus and Democritus, became Being, and a line of demarcation was thus drawn between spirit and matter, the infinite and the finite, and so on, which eventually condensed into the mind/body dualism and other such confusions. We have experienced successive fractionation of metaphysical doctrines and dogmas, conceptual frameworks and cosmologies, epistemological certainties and sophisms, all of which have led inexorably toward increasingly finer distinctions, the conjunction of part or all of which could be none other than inconsistent. But obstinacy prevails; and the search, nursed by feeble hopes, tends to surface occasionally.

In the present chapter, I extend the discussion of paradoxes, specifically, those of infinity, in order to present Cantor's

Chapter Three

puzzling theory of sets as well as Gödel's proof—that bête noire of mathematicians—and their relation to certain of Borges's texts. I then conclude on a contemporary note: the imminence of an ebb tide, if not an out-and-out demise, of those smug assurances in every field of intellectual endeavor.

> [T]here was an affinity with the ancient Greek contradiction about Epimenides the Cretan, who said that all Cretans are liars. It seemed unworthy of a grown man to spend his time on such trivialities, but what was I to do?
> —Bertrand Russell

Models of Infinity

1 The Greek word *apeiron* has been consistently translated "infinity." The original idea behind it, however, was not infinity in the modern mathematical usage but something more akin to that which is inconceivable, or chaos. The Greeks' fear of infinity was perhaps equaled only by their fear of irrational numbers. With the Renaissance concept of *coincidentia oppositorum,* the infinite became explicit in the finite and the finite implicit in the infinite. The fear had by and large subsided, and infinity was looked upon somewhat like an intuitive concept. In the seventeenth century, however, anguish returned. It was now believed that number could only be predicated of the finite, for infinity belonged exclusively to God. Infinity was therefore also beyond determination, since, if determined, the Absolute could no longer be regarded as infinite, but necessarily finite by definition. This renewed *horror infiniti* was especially evident in Blaise Pascal, whose work, Borges tells us,

> does not project the image of a doctrine or a dialectical process but of a poet lost in time and space. In time, because if the future and the past are infinite, there will not really be a when; in space, because if every being is equidistant from the infinite and the infinitesimal [according to "Pascal's Sphere" (*OI,* 6–9)], there will not be a where." (*OI,* 93)

Today, following many heated debates, certain periods of monumental disappointment, and even some cause for rejoicing, infinity of one form or another is generally embraced. Sur-

prisingly, it has even been appropriated for use in mathematical descriptions of "reality."

Borges's writings bear on many contemporary problems and aborted solutions to the paradoxes of infinity. Ana María Barrenechea (1965, 43) remarks, perhaps correctly, on the anxiety the Argentine writer feels over the infinite "because of his vision of a universe incomprehensible to and untenable for the human mind." Perhaps also, Borges is much too playful for such "anxiety," though he once confessed: "One concept corrupts and confuses the others. I am not speaking of the Evil whose limited sphere is ethics; I am speaking of the infinite" (*OI,* 109). More appropriate, I believe, and as Barrenechea (1965, 23) states earlier, infinity is for Borges perplexing, but he is nonetheless attracted to it. Interestingly, Borges (*OI,* 109) tells us that he once wanted to compile a history of infinity. Why he did not do so only he could know, but most likely the bewilderment caused by the concept throughout history served to halt his efforts. This bewilderment is revealed in Borges's observation that the paradoxes of infinity constitute "one of God's perfections in Theology, a cause for argument in Metaphysics, as well as a popular point of emphasis in Literature, a revived abstract concept in Mathematics . . . and a true intuition on looking at the sky" (*A,* 164). Indeed, without infinity, and all such fictions, without being able to measure "reality" in terms of a fictitious absolute, perhaps we could no longer continue to think, in the contemporary sense of the term.

Bernard Bolzano (1950, 132–33) offers a truism: it is absolutely impossible for our imagination to picture the infinite. But thereafter he more appropriately tells us that it is "enough for our *intellect* to grasp it and recognize it as something that cannot be otherwise than it is." I believe Borges would agree. He evidently discards the idea that infinity is "very, very large." Infinity, it was pointed out in the previous chapter, is of two sorts: *actual* and *potential.* Infinity can also be either denumerable (and potential) or nondenumerable (and actual). In brief, a denumerable infinite is that of the whole integers $(1, 2, 3, \ldots n)$. The nondenumerable infinite, complex numbers being a prime example, consists of a collection of entities along a line such that between any two there is an infinity of others. Such an infinity of points is a *dense continuum.* The example I have given for a denumerable infinite is *nondense, discrete.* However, although it can theoretically be "counted," the end can never actually be reached: it is ordinarily conceived as an interminable

Chapter Three

potential. The example for the dense nondenumerable infinite is, of course, inconceivably fine, smoother than the richest vanilla ice cream.

In another manner of speaking, a line is conceived by the infinitist to be actually infinite, but by the finitist to represent potentially an infinite number of points. The infinitist can take a sheet of paper, bisect it with a straight line, and confidently proclaim to have constructed infinity; the finitist finds a larger piece of paper, bisects it, and responds dryly that the first line was merely incomplete, for his line contains more points. The infinitist retorts that both contain the same number of points, for between every pair of points there is an infinity of others. The paradoxes of infinity are derived precisely from this second example.[1]

Our initial and naive impression of matter is invariably that of something divisible into infinity, and ever so small a part of it appears to us to possess the same properties as the whole. This is necessarily an artificial concept, for we must eventually admit that nature is "grainy." Yet we seem to persist in our occasional effort to reduce the furniture of our world to our ordinary conception of time and space. This propensity to mutilate the continuum with discreteness, or to interject continuity into discontinuity, is relevant to the Achilles and tortoise argument. Achilles never overtakes the tortoise because the series can never end. The infinitist declares that there is no problem, however, for infinity is actual; Achilles has merely to run through the infinity of points along a line. However, since Zeno's paradoxes are *predicated* on the concept of a *potential infinity*, finitistic arguments, to repeat, cannot legitimately dispel him.

Borges, sometimes a finitist, begins "The Book of Sand" (*BS,* 117) on a note of irony:

> The line is made up of an infinite number of points; the plane of an infinite number of lines; the volume of an infinite number of planes; the hypervolume of an infinite number of volumes. . . . No, unquestionably this is not—*more geometrico*—the best way of beginning my story. To claim that it is true is nowadays the convention of every made up story. Mine, however, *is* true.

Here we have enacted, once again, the rivalry between the finitists and the infinitists. The former would declare the first

sentence false; the latter, that is true. Borges subtly denies that this is a fitting manner to begin his story, for all fictions (i.e., mathematical, scientific, philosophical, religious, and even literary) are conventionally considered to be true, but they are invariably over the long haul not true. Even in mathematics, the most rigorous of disciplines, yesterday's verity can become today's nonsense, and vice versa. Borges's story, however, *is* true, or so he tells us. The important point is that between Borges's fiction and the mathematician's line representing an imaginary and infinitesimally thin division between one entity and another, there is a fundamental difference. The mathematical fiction has nothing necessarily to do with the experienced world; Borges's story, in contrast, is populated with the things of everyday living. The former can be looked upon passively, such as a straight line drawn down a sheet of paper. The reader cannot help but become involved in the second.

The Demise of Totalizing Quests

"The Book of Sand" (*BS,* 117–22) evinces a remarkable model of the discrete, denumerable infinite. One day an extraordinary stranger carrying a strange book visits Borges. On inspecting it, Borges discovers that it is ordered in versicles, like the Bible:

> In the upper corners of the pages were Arabic numbers. I noticed that one left-hand page bore the number (let us say) 40,514 and the facing right-hand page 999. I turned the leaf; it was numbered with eight digits. It also bore a small illustration, like the kind used in dictionaries—an anchor drawn with pen and ink, as if by a schoolboy's clumsy hand. (*BS,* 118)

The stranger asked Borges to locate the first page. He put his thumb in the flyleaf and attempted to open it, but it was useless. Every time he tried, a number of pages came between the cover and his thumb. It was "as if they kept growing from the book." His effort to find the last page was equally futile. It could not be, but it was. The book had an infinity of pages. Borges came into possession of the book, and during the months that followed he arose at three or four in the morning to leaf through its pages; he had become a prisoner of it. Finally, he realized that "the book was monstrous. What good did it do me to think that I, who looked upon the volume with my eyes, who held it in my hands, was any less monstrous? I felt that the book was a nightmarish object, an obscene thing that affronted and tainted reality itself" (*BS,* 121–22).

Chapter Three

There is a counterpart to this book mentioned in "The Library of Babel." It is a single volume consisting of an infinite number of infinitely thin leaves with an "inconceivable middle page" that has no reverse, and it is capable of replacing all the books in the interminable Library. This middle page of the infinite book can have no reverse since, if the book contains all possible sentences and their countersentences, such a page must be the mathematical equivalent of zero. And, like zero, that mysterious "nonnumber," it is at once the center and no-thing—unintelligible. Just as there is no negative zero, so there can be no opposite to the central page. This central page reminds one of the Buddhist concept of *Sunyātā,* the void or emptiness, a word also derived from "cipher" or "zero" in Sanskrit. Zero is at once no-thing and it "contains" the possibility for everything. That is, it is not merely some-thing empty, but rather, it is dynamically the potential for the generation of all integers. In the same way as zero, *Sunyātā* does not mean complete nothingness, it has both negative and positive facets simultaneously. Hence the void or zero, like *Sunyātā,* is unthinkable.

There is an interesting alternative to Borges's book, proposed by José Benardete (1964, 236–37). Suppose a book is lying on a table. Open it. Look at the first page. Measure its thickness. It is very thick indeed for a single sheet of paper—one-half inch thick. Now turn to the second page. How thick is it? One-fourth inch. And the third page is one-eighth inch, and so on. We are to suppose not only that each page of the book after page one is followed by an immediately preceding page, but also (and this is not unimportant) that each page is separated from others by a finite number of pages. Now close the book and turn it over so that the front cover is lying face down upon the table. Very slowly lift the back cover with the aim of exposing to view the infinite stack of pages lying beneath it. *There is nothing to see!* There can be no last page in the book to meet our gaze! Strange as it seems, this image, or un-image as it were, is also comparable to a model of the universe currently used by some mathematicians: the torus, whose center is nothingness, or a void, like *Sunyātā.*[2]

The Book of Sand, the book containing the infinite (or perhaps quasi-infinite) Library, and Benardete's variation, illustrate a most important point. They are all tropologues, mental constructs *representing* the actual infinite, but, to use Borges's phrase, they "affront and taint" the idea of the potentially infinite. The two types of infinity are mutually exclusive, of course.

The Demise of Totalizing Quests

And this is precisely the problem with Zeno. His paradoxes are about time, yet his mental construct presupposes a timeless state of existence. That is, within time, and commensurate with the notion of potential infinity, generativity—of words, numbers, increments, etc.—can be enacted in a linear sequence. Not so, however, regarding the actual infinite. Borges's infinite Library (if it is indeed infinite) is, alas, condemned to time.[3]

The harmony of the universe conceived from the perspective of a totality outside time, like the actual infinite, knows only the tropological equivalent of one musical form—the *legato*. The symphony of number, or discrete series, knows only its opposite—the *staccato*. All attempts to reconcile this dichotomy are based on the hope that an accelerated *staccato* may appear to our senses as a *legato* (Dantzig 1930, 169). With a Zenoesque argument, and by artificially mixing *legato* with *staccato,* it can easily be proved that Hladik could not have been executed. Suppose one half of a minute before the instant he was to be shot, A, his mortal enemy decided to kill him; but three quarters of a minute before the fatal moment, B, a jealous lover, is prepared to take revenge; seventh eighths of a minute before, C lies in wait for him with his rifle; at fifteen sixteens of a minute, D, a psychopath, plans to end Hladik's life; then E, then F, and so on. Since this continuation has no end, there can be no beginning of the minute before the execution; therefore Hladik could not have died. He could very easily have survived the next infinity of years, meanwhile completing an infinity of plays!

Must such apparent madness be taken seriously? No, Borges would say; it is only an amusing game, but with a purpose: to illustrate the fallacy of all dogmas, all metaphysical doctrines. Such paradoxes reveal that even the most abstract reasoning can never free us of ambiguity, of fuzziness, of uncertainty. Benardete (1964, 253) calls it "sludge." There will always be

> sludge that frustrates our efforts to implement the ideal in practice, not only in the present case but on all occasions. sludge! . . . that unruly, refractory element in the world that, assuming the form of mechanical friction in one case, renders the perpetual motion machine impossible and that, assuming the form of Gödel's theorem in another, renders the consistency of mathematics incapable of proof. Sludge is everywhere, it cannot be escaped, not only in the physical world but even in the realm of pure ideas. Although this principle of recalcitrance is familiar to all, it has been generally believed to be merely contingent in nature, a brute fact that has no intelligible warrant.

Chapter Three

Georg Cantor, whose work Borges knew well, paid no attention to such disparities between our ideals and our real capacities, our "mental-realities" and our "sensory-realities," and partly as a result of his efforts, paradoxes of the Zeno variety enjoyed a rebirth of interest at the end of the nineteenth century. This most controversial mathematician argued not only that infinite sets are real but also that they obey their own unique laws. Cantor referred to these sets as "transfinite numbers," choosing to represent them by the first letter of the Hebrew alphabet, Aleph, with the symbol, $\aleph_0$, pronounced "Aleph-null." The subscript distinguishes $\aleph_0$ from other "larger" infinities Cantor "discovered," which he denoted, $\aleph_1$, $\aleph_3$, $\aleph_2$, and so on. Cantor was apparently startled to find not only that there are sets "larger" than $\aleph_0$ but also that there are many infinite sets equal to $\aleph_0$. For example, we apparently assume intuitively that there are fewer even numbers than there are whole numbers, but actually the magnitude is the same. Cantor demonstrated this with one-to-one correspondences between infinite sets:

$$\begin{array}{cccccc} 1 & 2 & 3 & 4 & 5 & \ldots n \\ \updownarrow & \updownarrow & \updownarrow & \updownarrow & \updownarrow & \\ 2 & 4 & 6 & 8 & 10 & \ldots n \end{array}$$

The same can be constructed for odd numbers or any other ascending combination of numbers, and identical results obtain. On the basis of such correspondences, Cantor declared that, though it is theoretically possible to write down the infinity of whole integers (a discrete series), it is altogether impossible to write down all the real numbers, i.e., π, $\sqrt{2}$, etc. What was more confounding yet, Cantor "discovered" that the addition and multiplication of Aleph-null does not follow standard laws of arithmetic. That is,

$$\aleph_0 + \aleph_0 = \aleph_0, \text{ or } 2 \times \aleph_0 = \aleph_0$$

Cantor's strange arithmetic is made possible by the unique laws of sets. In the "new mathematics," children are taught at a tender age to use set theory. A set of five hats combined with a set of seven shirts forms a set consisting of two sets with a total of twelve elements. This is simple enough. However, since the set of all whole integers is equal to the set of all even integers, which is $\aleph_0$, their compound set must also be $\aleph_0$. And yet com-

mon sense tells us that the set of even integers can be no more than a part of the set of whole integers. By the same token, an increment of a line contains an infinity of points just as does the entire line, so it is equal to the line, yet it is no more than a part of the whole. No matter to what extent we expand an infinite set, or into however many points we divide it, all of them will be equal. This seemed to be second nature to Cantor, who, in his own words, was

> so in favor of the actual infinite that instead of admitting that Nature abhors it, as is commonly held, I hold that Nature makes fervent use of it everywhere, in order to show more effectively the perfections of its Author. Thus I believe that there is no point of matter which is not—I do not say divisible—but actually divisible; and consequently the least particle ought to be considered as a world full of an infinity of different creations. (quoted in Dauben 1979, 124)

Everything is contained within everything else![4]

There is an obvious circularity in attempting to think of the set of all sets, for try as one may, one finds it impossible to regard it as a completed, self-contained entity. Yet Cantor believed his proof demonstrated that whatever the magnitude of a given set may be, it is always possible to construct one of greater magnitude. And what is more perplexing, he proclaimed these infinite sets to be real (Dauben 1979, 242).[5]

Borges was especially fascinated during his writing of *Ficciones* and *El Aleph* with the notion of a set that contains itself. His discussion of set theory in *Historia de la eternidad* and his review of *Mathematics and the Imagination* (*D,* 165–66) by Edward Kasner and James Newman reveal his familiarity with Cantor. Borges appreciated the fact that Cantor's work led directly to the paradoxes of infinity and self-reference. His narrator's experience of the Aleph, the entire universe contained within a minuscule part of itself, as the "only place on earth where all places are—seen from every angle, each standing clear, without confusion or blending" (*A,* 23), definitely implies Cantor sets.[6] There is a crucial distinction, however, between Cantor's Alephs and Borges's Aleph. Since Cantor believed them to be real, he never relinquished hope that the paradoxes they generated would be ultimately resolvable. Borges, of course, believed in the existence of no such eternal archetypes.

Chapter Three

The Alephs interested him not because he thought they were real but because they allowed him to demonstrate that nothing is real, except the relations making up the universe (Hayles 1984, 159–61). Borges thus undercuts Cantor's vain expectation that infinite sets can be found within a logical system of thought. The irony is that Cantor brought about the same transformation of perspective in his treatment of the Alephs.[7]

In addition to "The Aleph," "The House of Asterion" (*L*, 138–40) evinces an excellent example of the Cantorian image. The house, inhabited by the Minotaur, is a labyrinth with an infinity of doors—which for the Minotaur is, surprisingly, equal to fourteen! The Minotaur never leaves the house—though he confesses with some pride that he once ventured out into the city for a short while. The fourteen doors are not locked, he claims, so he actually is not a prisoner; both men and animals can come and leave as they wish. Assuming these doors to exist in succession—and we have no reason to believe otherwise, since the house is a labyrinth—either he cannot leave or the outsiders cannot enter.

Ironically, the fusion of infinity with finity (the number fourteen) leads to a bizarre Zenoesque quandary. If one wishes to pass through the doors, one must open door fourteen, then one must open half of that number, door seven, then, strangely, door three and one-half, and so on (but perhaps this is not so strange, since if the fourteen doors are equal to infinity, in Cantorian fashion, then between each pair of doors there is an infinity of "doors"). In contrast, if one decides to enter from the other end, *there can be no final door;* in fact, just as Benardete's variation of Borges's Book has no last page, so there is nothing to be perceived from the other end of the "infinite" series of fourteen doors. Obviously, the Minotaur stands before door number one, for he knows a door is there. Hence the outsiders have no barrier preventing their passage into his domain; from their side, there is simply no door to be seen. In contrast, the Minotaur, in spite of himself, is trapped within his idealistic prison, for he will never be able to pass through the "infinity" of doors before him. Of course, this paradox holds if and only if this strange form of infinity is considered incomplete rather than Cantorian. Borges has once again elusively crossed incommensurable barriers to confound us.

The house's sole inhabitant also presumptuously announces that the art of writing is useless. In fact, since the Mino-

taur never bothered to learn to write, he cannot retain the difference between one letter and another. This prepares us for his proclamation: "All parts of the house are repeated many times, any place is another place. There is no one pool, courtyard, drinking trough, manger; the mangers, drinking troughs, courtyards, pools are fourteen (infinite) in number. The house is the same size as the world, or rather, it is the world" (*L*, 139).

The deck of cards has been switched on us once again, for we seem to be back in the same framework as Cantor. Each part of the house, though infinitesimal, is repeated infinitely, and each part is, paradoxically, equal to the whole. Further, in the Minotaur's mind at least, the writing he abhors possesses the same property. We must believe this to be so, for it is implied that the Minotaur is an inveterate idealist: "[T]wo things in the world seem to be only once: above, the intricate sun; below, Asterion. Perhaps I have created the stars and the sun and this enormous house, but I no longer remember" (*L*, 140).

Russell, in his attempt to clear up what most mathematicians considered to be Cantor's muddle, constructed his "Theory of Logical Types." He demonstrated, as mentioned above in the discussion on "The Circular Ruins," that a set of things cannot itself be one of those things, e.g., the class of all women is not a woman, and that the set of all sets cannot be a member of itself. To illustrate this problem, let us further consider Borges's "Library of Babel." The inhabitants of the Library hold a superstition about a so-called "Man of the Book." There is supposedly a book that is "the formula and perfect compendium *of all the rest:* some librarian has gone through it and he is analogous to a god" (*L*, 56). But how could one hope to locate the unique hexagon housing this book? A regressive method was proposed: "To locate book A consult first a book B which locates A's position; to locate book B, consult first a book C, and so on to infinity" (*L*, 56). The narrator confesses that he has "squandered and wasted" his years in such adventures, and now it seems unlikely to him that "there is a total book on some shelf of the universe" (*L*, 56–57).

Curiously, Borges reveals in a footnote that "[I]t suffices that a book be possible for it to exist. Only the impossible is excluded." For example, "no book can be a ladder, although no doubt there are books which discuss and negate and demonstrate that possibility and others whose structure corresponds to that of a ladder" (*L*, 57). Assuming the Book of Books to be a

The Demise of Totalizing Quests

possibility, then, it must somehow exist—a moderate concession to Cantor. Now the problem that remains is this: the Book is a compendium of all the books, but presumably it is not included among them. This saves the book from at least one contradiction, for if it were to be included among the other books, it would necessarily include itself in the compendium, and Russell's paradox would be in effect. Yet the compendium is, itself, a book included in the Library. To be complete, and we must suppose that it is, the self-contained Library must contain another compendium containing both the first compendium and all the other books—and an infinite regress begins. In essence, what occurs here is related to Cantor's Alephs. Assuming that all the books in the Library less the compendium are equal to infinity, when the new compendium is written, it will contain the original infinity of books plus one, which is still infinity, and when yet a new compendium is written, one book must be added to the second infinity of books, which is still infinity, and so on. The Total Library under these conditions escapes Russell's paradox, since it supposedly does not contain itself, but it cannot escape the infinite regress.

The Library, in this sense, is reminiscent of David Hilbert's Grand Hotel, which he used to illustrate the work of Cantor and other paradox-mongers. Imagine a hotel with an infinity of rooms, all of which are full of guests. A person arrives in need of a room. "Of course," replies the manager, "we will be pleased to accommodate you." "But sir," sheepishly pleads a nearby bellhop, "I believe all the rooms are occupied." "That's no problem," the manager exclaims and proceeds to move the person occupying room one into room two, the person occupying room two into room three, and so on. Then, finally, the new arrival is free to deposit his bags in room one. Now suppose an infinite number of new guests arrives. Still no problem. The manager simply puts the first infinity of guests in the even numbered rooms and the second infinity in the odd numbered rooms. The hotel is inexhaustible!

One of the more interesting appropriations of Cantor's arithmetic, related also to Hilbert's paradox, comes from the Chilean psychoanalyst Ignacio Matte Blanco (1975), who divides the mind into what he calls "symmetrical (unconscious) being" and "asymmetrical (conscious) being."[8] Consciousness and thinking are linear and entail successively taking one thing after another (analogous to the Zahir). However, Matte Blanco (1975, 98) suggests: "Nothing prevents us from conceiving a

form of consciousness which can apprehend an infinite number of things simultaneously" (analogous to the Aleph). This is symmetrical being in the undivided state before it has been severed into particular states of consciousness. It entails a potential infinity of sets: being without happening, spaceless and timeless. Asymmetrical being, in contrast, must apprehend Hilbert's hotel rooms in succession; symmetrical being is capable of mentally grasping the entire set of rooms in simultaneity. Perhaps, says the author in all seriousness, symmetrical being corresponds to the consciousness of God!

Any attempt to comprehend Matte Blanco's symmetrical being rationally is futile. At most, it is perhaps accessible to mystical experience, which is dear to Borges and perhaps most adequately illustrated in "The God's script" (*L*, 169–73), to which I now briefly turn in order to relate it to Matte Blanco's concept. Tzinacán, an Aztec priest, was captured and tortured by Pedro de Alvarado and thrown into a dome-shaped dungeon divided by a wall bisecting it, with him in one half and a jaguar in the other. He was, we are told, the last priest of the god who had written a secret formula in an unknown place for the last man to read. This formula, with which Tzinacán became obsessed, could supposedly give omnipotence and immortality to its decipherer. After deciding that the sacred script lies concealed in the jaguar's spots, Tzinacán devoted years to learning their configuration in order to decipher the inconceivable text. He once asked himself what type of sentence an absolute mind would construct, but, try as he might, he could not discover an answer to this enigma. Finally

> I considered that even in the human languages there is no proposition that does not imply the entire universe; to say *the tiger* is to say the tigers that begot it, the deer and turtles devoured by it, the grass in which the deer fed, the earth that was mother to the grass, the heaven that gave birth to the earth. I considered that in the language of a god every word would enunciate that infinite concatenation of facts, and not in an implicit but in an explicit manner, and not progressively but instantaneously. In time, the notion of a divine sentence seemed puerile or blasphemous. A god, I reflected, ought to utter only a single word and in that word absolute fullness. No word uttered by him can be inferior to the universe or less than the sum total of time. (*L*, 171)

In Tzinacán's conception of things, the sentence, the consecrated word itself, has become the totalizing, timeless Word,

which is contained within the entire universe, which is *it*—Cantor, once again. Possession of this Word in all its plenitude, however, is beyond him. Enshrouded in darkness, days and nights become the same, one dream tends to become all dreams—timeless symmetry. Yet there is a linear, asymmetrical progression. Tzinacán one night dreams, and upon awakening there is a grain of sand on the floor; he sleeps again, and finds a second grain of sand when he wakes up. He continues on until he lies buried beneath a hemisphere of sand. He tries in vain to arouse himself and is told: *"You have not awakened to wakefulness, but to a previous dream. This dream is enclosed within another, and so on to infinity, which is the number of grains of sand. The path you must retrace is interminable and you will die before you ever really awake"* (L, 172).

There are two paradoxical images reiterated here. The first is encapsulated in the proposition "All dreams are one dream"—the tradition of Schopenhauer and the Upanishads. This may be called, in Cantorian fashion, the "hologram model."⁹ The second paradox is tantamount to the proposition "Each dream is enclosed within another dream." This is the "Chinese-box model."¹⁰ The two images are necessarily incommensurable, since, according to the first, which is symmetrical, each dream contains all (an infinite number of) dreams and all dreams contain each dream; and according to the second, an asymmetrical formulation, each dream can contain only the dreams it contains and it in turn is contained within an infinite number of successive dreams. The first, in Matte Blanco's conception, is symmetrical being's grasp of the rooms in Hilbert's Hotel in an instant—to say "one room" would be to say "all rooms." The second implies asymmetrical being's inclusion of an incomplete set of particulars within its limited conceptual grasp, which is superseded (metaphorically) or displaced (metonymically) by a successive grasp, by all successive grasps.

It has become apparent that Borges's ludic proclivity evokes simultaneously fascination and incredulity regarding Cantor sets and trust in either an actual or a potential infinite. We shall note, in the following section, that Borges promises even more: our immanence, within a self-reflexive, self-contained, quasi-infinitely recursive universe of our own making, prevents any unitary grasp of the Totality. We are inextricably condemned to our finitude.

Mathematics is a human activity and is subject to all the foibles and frailties of humans. Any formal, logical account is a pseudo-mathematics, a fiction, even a legend, despite the element of reason.

—Morris Kline

The Demise of Totalizing Quests

2 At the turn of the century, Hilbert and other mathematicians envisioned a total axiomatization of mathematics by which proof of the consistency of any and all mathematical systems could be realized. Such quests for the absolute came to an abrupt halt, however, when in 1931 Kurt Gödel published his earthshaking and in many circles unwanted theorems in "On Formally Undecidable Propositions of *Principia Mathematica* and Related Systems." Gödel demonstrated that, given any logico-mathematical system strong enough to express the arithmetic of natural numbers, either (1) the ultimate truth-value of the formal system cannot be determined from its own set of axioms but only from some outside axiom (the "incompleteness principle"), or (2) the system cannot be totally free of hidden contradictions (the "inconsistency principle").[11] Roughly speaking, Gödel's proof is comparable to the paradox of the Cretan who declared that all Cretans were liars. If he was telling the truth, he himself was lying, and if lying, he was telling the truth, and so on. As a result of Gödel's and other comparable discoveries, it is now admitted (although at times reluctantly) that there can exist no single, closed, consistent formal system capable of including the totality of logic and/or mathematics. All systems are inexorably incomplete, since their root axioms, if consistent, can only be proved from without. There can be no absolute origin or center.

In the aftermath of the Gödelian crisis, most mathematicians have become resigned to the fact that their discipline can no longer be considered a bastion of certainty. But this has been no cause for alarm. As Miguel de Unamuno (1954, 109) tells us, "the supreme triumph of reason is to cast doubt on its own validity."[12] Significantly, Morris Kline, an excellent mathematician in his own right, argues that the virtue of a proof is not that it compels belief but that it suggests doubt; it calls for disproof and tells the mathematician where to concentrate his doubt.

The End of Certainty

Chapter Three

Kline (1980, 316) concludes: "Mathematicians have been worshipping a golden calf—rigorous, universally accepted proof, true in all possible worlds—in the belief that it was God. They now realize it was a false god." Mathematics, commensurate with Wittgenstein's normative concept outlined in Chapter Two, is no longer considered to be immutable. It grows not by the increase of indubitably established theorems but by the improvement of guesses, speculation, and criticism. The royal road to the Absolute is now clearly blocked.

It is axiomatic that a "loss of certainty" also began in the arts shortly after the turn of the century. It was not, however, until around the time of Gödel's proof that prose fiction became hyperconsciously self-referential. Today, readers of Beckett, Ionesco, Pirandello, Vladimir Nabokov, Alain Robbe-Grillet, Natalie Sarraute, Borges, and others experience a situation comparable to that of the mathematicians when Gödel demolished their most cherished dreams. Much of these authors' narrative, like mathematics according to Gödel's proof, is self-negating, recursive, and infinitely regressive.[13] Of course, Borges was to a degree a child of his times, as are all authors, but he was also a precursor in many respects. I have discovered no indication that he had read anything on Gödel before writing *Ficciones* and *El Aleph*. Yet, as we shall observe, he was most likely intuitively aware of what have become the implications of undecidability of the "Cretan liar" type. As an example of this awareness, the Book of Books in Borges's Library is set apart from all other books, so it apparently does not speak of itself and is not self-referential. The chief problem with the Library system regarding the existence of this Book is that the Library will be eternally incomplete; that is, to account totally for the Library plus the Book demands the existence of another more all-encompassing Book, and then another, and so on. Therefore, as pointed out above, an infinite regress is implied.

In order to substantiate this infinite regress properly and to relate Gödel to Borges's Library, let us briefly outline Gödel's strategy. First, assign a unique index number to each finite formula in the mathematical system: these may be called "code numbers." Then, assign a unique index number to each proof in the system: these may be termed "proof numbers." Now suppose it is possible to prove that *within the system itself* there exists, say, the following sentences: (1) "The formula '5 + 5 = 10' has the code number 777," and (2) "The proof number of the formula with code number 777 is 851." This appears to be innocent

enough, a trivial game of labeling. The sentences are easily proved to be either true or false merely by matching the code number with a particular formula in the system and by matching the proof number with a particular code number—but notice that the code number and the proof number exist at distinct levels. Gödel, by a crafty operation, demonstrates the equivalent of saying that an undesirable formula exists somewhere in the system with a certain code number, say, 2,619, which states: "A proof number for the formula with code number 2,619 does not exist." This is called a "Gödelian sentence." In other words, when the code number 2,619 is assigned to a formula, there can be found no proof number of the formula with code number 2,619 from within the system, and hence the system is unprovable in and of itself. It can be rendered provable solely by generating an axiom from outside the system. So the system is either *incomplete* (its proof requires an outside axiom) or *inconsistent* (it contains the equivalent of a sentence that states its own falsity). The relevance of the Cretan liar paradox to the Gödelian sentence now becomes evident.

In a similar vein, reconsider Hofstadter's "authorship triangle." As a variation of this scheme, and more commensurable with Borges's aesthetics, let us call it a "dream triangle." There are three dreamers, X, Y, and Z. X exists in Y's dream, Y in Z's dream, and, naturally, Z exists in X's dream. This system as it stands is self-reflexive and inconsistent. But there *is* a way out, provisionally at least: X, Y, and Z exist in O's dream. From within the "tangled loop" there can be no more than recursiveness to infinity. However, O, existing outside the space of X-Y-Z, knows of their existence, but they are oblivious to his. In order to represent O we must, so to speak, jump off the two-dimensional page.

Discourse can also exist within a given frame, and, like O's dream of X-Y-Z, meta-discourse *about* that frame can be generated from "outside." The problem is that one's successively stepping outside frames of discourse potentially creates a series the totality of which is available solely to an infinite perspective, something like the hypothetical author of the Book of Books in Borges's Library. Finite beings that we are, we can be in possession of no such perspective. The task then, is to discover an explanatory mechanism with which to account for an indefinitely expanding recursive series *from within*. Back to Borges.

The narrator of "The Library of Babel" tells us at one juncture that the Library might *appear* to be infinite. However,

Chapter Three

since it contains all the possible permutations of the twenty-five orthographic symbols, and even though the total number of books must be monstrous, nonetheless the Library is necessarily finite. The solution to this quandary, suggests the narrator, stems from the fact that a finite system can apparently become infinite if it is infinitely perpetuated. That is, after all the permutations are exhausted, then they repeat themselves limitlessly and hence become, rather incomprehensibly, a finite but unbounded system. This premise of the eternal return, we are told at the end of Borges's story, stems from the effort to order the disordered, the compulsion to find meaning in the universe (i.e., the Grand Unified Theory). Interestingly, an impious group exists which maintains that "nonsense is normal in the Library and that the reasonable (and even humble and pure coherence) is an almost miraculous exception" (*L*, 57). But the narrator persists in his search for harmony amidst the apparent chaos. He prays to the unknown gods "that a man—just one, even though it were thousands of years ago!—may have examined and read it [the Book of Books]" (*L*, 57). He longs for the instant of enlightenment to be able to apprehend the Order: "let Your enormous Library be justified" (*L*, 57). But the narrator is never allowed this moment of insight. The most he can do is "venture to suggest this solution to the ancient problem: '*The Library is unlimited and cyclical*' " (*L*, 58).

Why is the narrator condemned, like the fabled Minotaur, to remain sealed within his prison house (of language)? A response is problematic, but perhaps it can be initially approached as follows. First, Borges writes the text, and the narrator narrates it, but Borges *is* the narrator, so the text includes Borges—which is not so strange, since he has declared that Don Quixote is the reader of the *Quixote,* and Hamlet is the spectator of *Hamlet.* Second, the reader reads (includes, assimilates) the text, and since Borges is the narrator narrating it, therefore the reader "reads" (includes, assimilates) Borges—the "other Borges," that is to say. However, we must contend with the Library, which is ominous indeed, for the Library presumably includes all. The narrator, we discover, is to be found within the Library, writing the manuscript of "The Library of Babel" with his "fallible hand" wherever he can, on the "cover of a book, with the organic letters inside: punctual, delicate, perfectly black, inimitably symmetrical" (*L*, 53). But we are obviously *not* reading this original manuscript, as Hayles (1984, 151–52) remarks, but

a book that is now stored in the Library. If so, then we the readers, as well as the narrator (Borges), are actually included within the Library.

The Demise of Totalizing Quests

In other words, the Library, whether infinite or finite but unbounded, must be either inexorably and self-reflexively *inconsistent* or *incomplete*. That is to say, the story speaks of the Library, which contains the story as well as its narrator (Borges) and its readers. As such, the Library contains the story, which in turn contains the Library, paradoxically. The Library, presumably for the readers of the story, is fictive, and the story, contained within it, is also a fiction. It is as if the Library, through the story that it contains, were addressing itself to the reader, who believes herself to be outside the library but in reality is not, with the proposition, "This is *not* a fiction"—the subject of the proposition being either the Library or the story, since they contain each other. In similar fashion, though not the equivalent, each dream of the X-Y-Z system implicitly puts forth the proposition "This is *not* a dream." A painting of a pipe that asserts implicitly (and explicitly in Magritte's work), "This is *not* a pipe," evinces a comparable dilemma. The inconsistency in each case is tantamount to the Gödelian sentence (or Cretan liar) which says of itself, "This sentence is *not* true." However, a sentence of the liar type, "I am lying," if contextualized, can be made properly consistent. Perhaps the utterer was actually referring to a previous utterance. Or perhaps the listener, upon questioning him about his utterance, obliges him to recant. In either case, clarification demands a step outside the sentence. So if the sentence is inconsistent, then it is falsely self-sufficient; and if it is made consistent, then it ultimately remains incomplete.

I have been speaking of the paradoxical Library as a Totality, as infinite. If indeed it is such a Totality, then it is doomed to its own inconsistency, for there can be no redeeming "axiom" from outside to justify it. On the other hand, if the Library is finite, as some of its inhabitants hope, then it is forever incomplete, for the foreign "axiom" struggling feebly to make it consistent (even intelligible) is precisely the story's reader, the ongoing stream of readers. But once a given reader is seduced into the Library, it is the end of the road, for, like Escher's hands drawing themselves or the dreamers mutually dreaming themselves, that reader is now within the Library, and is bound to remain within—the border between fiction and world having been ruptured. No matter by what clever means the reader

attempts to break out of this circle, she cannot: she is, inexorably, immanent. It is as if she were a flatlander on a two-dimensional sheet, unable to transcend her sphere of existence.

The impossibility of one's transcending a system that binds one is remarkably demonstrated in Escher's *Dragon*, to which Hofstadter (1979, 473–75) refers (figure 2). The dragon, existing on a two-dimensional sheet, makes a heroic effort to transcend its sphere of existence. It struggles against its limitation, cutting imaginary—and we must call them imaginary—slits in the paper, attempting to stick its head underneath a fold and through in order to bite its own tail from another dimension. We the observers, from outside, can perceive the futility of it all. In order to aid the wretched beast, we could

> tear it out of the book, fold it, cut holes in it, pass it through itself, and photograph the whole mess, so that it again becomes two-dimensional. And to that photograph, we could once again do the same trick. Each time, at the instant that it becomes two-dimensional—no matter how clearly we seem to have assimilated three dimensions inside two—it becomes vulnerable to being cut and folded again. (Hofstadter 1979, 474)

All intellectual systems, claims Hofstadter, suffer the same limitations: the insurmountable barriers incessantly before us. It is not coincidental that Gödel's and other comparable theorems are called "limitative theorems."

However, though we are led to believe that Borges's chimerical Book of Books is inaccessible, and that the eternal traveler in the Library is destined to wander incessantly in circles that never exactly coincide, one might suggest, while obstinately clinging to a ray of hope, that there is yet a way out of this morass, for the narrator tells us that

> a clarification of humanity's basic mysteries—the origin of the Library and of time—might be found. It is verisimilar that these grave mysteries could be explained in words: if the language of the philosophers is not sufficient, the multiform Library will have produced the multiform language required, with its vocabularies and grammars. (*L*, 55)

Alas, this "inordinate hope" eventually led to "excessive depression." The certitude persisted that some shelf in some hexagon contained priceless books divulging the secrets of the universe. But the search proved, in the final analysis, to be

The Demise of Totalizing Quests

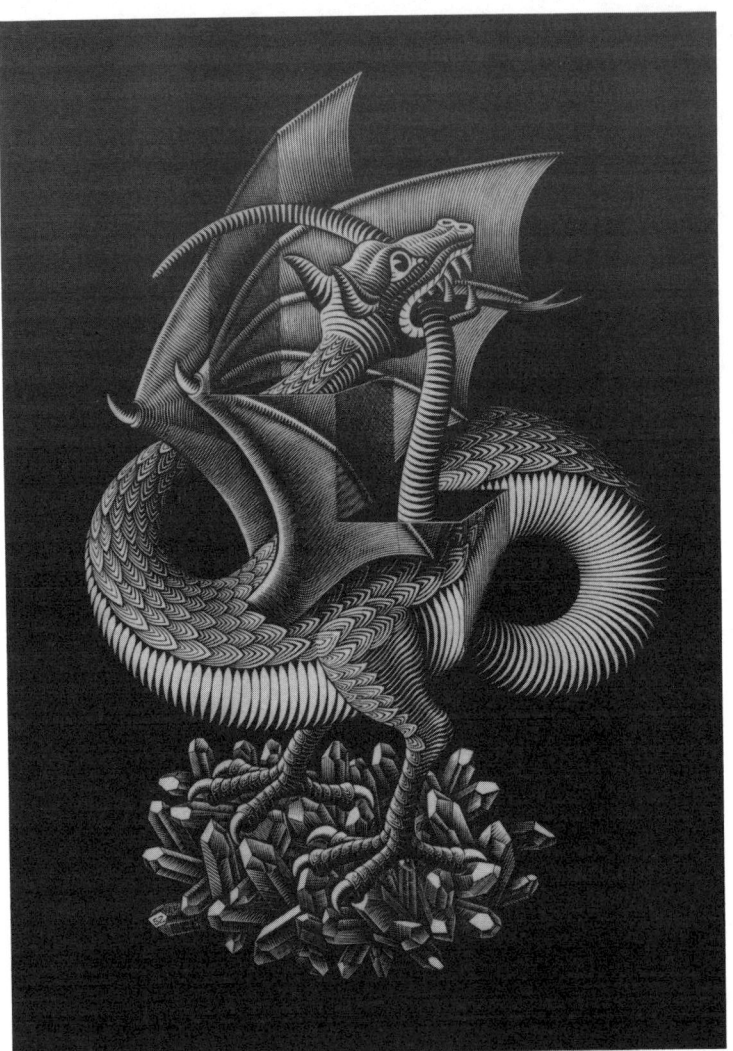

Dragon. 1952. © M.C. Escher/Cordon Art—Baarn—Holland

Figure 2

futile. Why? I would conjecture that it is due to the possible element of randomness that governs the Library. In a tangential flight, let us return to number in an effort to understand better the inhabitant's immanence within a random or quasi-random system.

Chaos, mythically the confused state of primordial material, is ordinarily understood not as randomness but in terms of irrationality. Consider, for instance, the decimal expansion of $\sqrt{2}$. Suppose that in succession any one of the digits from 1 to 9, and including zero, pops up at a given instant, and suppose that these digits correspond to different sounds that pierce our ears. It will appear chaotic. However, the fact remains that there exists a finite rule with which to generate the sequence of sounds, so the series is actually not chaotic. Now consider a real number, an infinite stream of digits with a finite rule by means of which to generate it—keeping in mind that the Library is for practical purposes infinite, but its inhabitants, being finite, require a finite rule for deciphering the totality. The decimal expansion of this number is presumably random, as opposed to the decimal expansion of a rational number, which exhibits a recurrent pattern. But to the question, "Is the number actually random?" an answer cannot be forthcoming. No matter to what length we generate the digits of this number, we still cannot say it is random, for, with the next digit that pops up we might discover some sort of order. Analogously, to the question, "Are the books in the Library random?", or conversely, "Does the Library manifest an inherent order?" there is no accessible answer. Yet, the travail is interminable; the narrator of "The Library" persists in his efforts to disclose the cosmic secret, the Order: "My solitude is gladdened by this elegant hope" (L, 58).

Discovery of the Library's order, if forthcoming, must be from within that order, and it necessarily entails a "bootstrap" operation. The finale of "Averroes' Search" (L, 148–55) elegantly illustrates the possibilities and problematics of such an operation, which I will treat in greater detail in Chapter Seven. Averroes, a renowned Arabic-Hispanic scholar who lived from 1126 to 1198, is at the outset of the story found at work in his library in Cordoba writing a refutation of a treatise by an antirationalist Persian. He is, however, distracted by a certain problem uppermost in his mind: a commentary on the works of Aristotle, especially regarding *tragedy* and *comedy*. His Islamic background, the absence of the theater, his language, all place him in a conceptual framework apparently incommensurable

with the Greek mind; there appears no hope of his making the two elusive terms adequately intelligible. He later honors an evening invitation at the home of Forach, a noted Koran scholar. A traveler, Abulcasim, is there also. He tells of his recent adventures in China, witnessing a strange event in a house where many people were sitting in rows, and some in balconies. This appears bizarre to Averroes and the others, for the theater does not lie within their range of cultural experiences. However, a spark of imagination persists in Averroes, and, while at home following the discussion, he solves his enigma: *tragedy* is "panegyric," *comedy* is "satire" and "anathemas," and in the venerable Koran can be found the illustration of both. Then, exhausted, he unwinds his turban and gazes at himself in a metal mirror, and at that moment he disappears, along with his house, the fountain, his books, the slave girls, and, we are told, perhaps the Guadalquivir River.

The narrator then interjects a postscript. Here, he reflects, is the case of a man who has set a goal for himself that is not forbidden to others but is to him:

> I remembered Averroes who, closed within the orb of Islam, could never know the meaning of the terms *tragedy* and *comedy*. . . . I felt the work was mocking me. I felt that Averroes, wanting to imagine what a drama is without ever having suspected what a theater is, was no more absurd than I, wanting to imagine Averroes with no other sources than a few fragments from Renan, Lane and Asín Palacios. I felt, on the last page, that my narration was a symbol of the man I was as I wrote it and that, in order to compose that narration, I had to be that man and, in order to be that man, I had to compose that narration, and so on to infinity. (The moment I cease to believe in him, "Averroes" disappears.) (*L,* 155)

The concepts of tragedy and comedy exist within the cultural milieu of the West, and hence the pair is for Borges adequately intelligible, but not for Averroes. Borges, on the other hand, endeavors to construct a narration that lies within Averroes's Islamic form of life, a task equally as impossible as that of Averroes. The self-reflective injunction both men give themselves is tantamount to the paradoxical Socratic knowledge paradox, which pragmatically puts one in an untenable situation, for to know *that* one knows, one must already know, and if one already knows, then one cannot conscientiously set out to obey the injunction. Yet, in a manner of speaking, both tasks *are*

possible, for Borges *does* complete his narrative, however inadequate he may claim it to be, and Averroes *did* somehow solve his problem, for his answer vaguely corresponds to Aristotle's *Poetics*. The coalescence of Borges and his narrative fictionalizes Borges in simultaneity with the realization of Averroes, but, in the final analysis, when Averroes discovers the key to his enigma, he fades back into oblivion, for it is at this point that Borges ceases to believe in his fiction. The upper level appears to have been somehow penetrated from within the circle, but at the same time it has not been reached because the actors remain within the circle. In this manner, the paradox has in a sense been resolved, yet it has not been truly resolved, since both Averroes and the narrator apparently merely muddled their way through to an answer; there was no way of their knowing absolutely how they stumbled upon it or whether or not it was correct.

Hence: Western World—Borges → Islam—Averroes → Western World—Borges . . . n consists of a recursive series that is, in a rough way of speaking, the cultural-linguistic counterpart to the above generation of an endless series of numbers—as well as the dreamers dreaming each other, and so on. Induction cannot determine the validity of the series, just as the Library's inhabitants cannot know by logical inference or mathematical formulae if the Library is totally disordered, nor can they determine any degree of order. This conjunction I have established between "Averroes" and "The Library" suggests that (1) there can be no rule or algorithm for determining the nature of a system within which one finds oneself, and (2) there is no guarantee that a given interpretation (e.g., of Greek drama for Averroes, of the Library for its inhabitants, of "Averroes' Search" for Borges, etc.) is valid—and hence fallibility prevails (significantly, the narrator of "The Library of Babel" *is* fallible).

Various alternative forms of the endless series I have commented on thus far are found in other works by Borges. His rarely studied essay "Time and J. W. Dunne" (*OI,* 18–21)[14] is about a strange scholar who argues in his "almost scandalous" books, especially in *An Experiment with Time* (1927) and *The Serial Universe* (1934), that dreams are derived from future as well as past events.[15] Dunne's work represents yet another aborted attempt to transcend our human limitations. His method, though he confesses that it "is not scientific evidence, nor is it intended to be regarded as such" (1927, 94), is nonetheless provocative, but not untried, even at the time of Dunne's writing. He and a large number of acquaintances recalled in the greatest detail possible

their dreams each morning and during the days that followed tried to discover their realization or partial realization in their everyday world. Dunne attempts further to illustrate the possibility of dream precognition with his theory of "serial time," which is something akin to an infinity of Chinese boxes. Every time-traveling field of presentation, i.e., an observer, is contained within a field one dimension larger, the larger field including events that are simultaneously "past," "future," and "present" to the smaller field. For example, our Time, T_1, is for us linear and trapped within three-dimensional space. To a four-dimensional observer, on the other hand, our time would be tantamount to another dimension of space at right angles to each of the three dimensions of our space, and this observer would see our past, present, and future in simultaneity. A fifth-dimensional observer would see the fourth-dimensional observer's universe in the same fashion, from yet a higher dimension, and so on, to infinity.[16]

Serialism of this field of presentation requires the existence of the human organism as a "serial observer" who is somehow mysteriously aware, though neither consciously or intentionally, of his own act of observation and who has another observer aware of his act, who has another observer, and so on.[17] In this manner we are not limited to three-dimensional space and one-dimensional time but transcend them at a higher level, at infinitely higher levels. We are, then, infinite; we exist in eternity—and death loses its sting. This is a most gratifying thought, except for the puzzle of arriving at the infinite dimension: What happens when we get there? Borges remarks, ironically, on concluding his essay on Dunne, that with "such a splendid thesis as that, any failing committed by the author becomes insignificant" (*OI*, 21).[18]

This elegant vision of immortality suggests another theme consisting roughly of an uncanny conjunction of Borges, Cantor, Dunne, and even recent speculations in quantum theory: the disintegration of the "I," or consciousness, as it were. Does the "I," conceived as an infinite Chinese box—*pace* Dunne—consist of a discrete succession, or does it become, ultimately, coterminous with all "egos"? Preoccupation with the "I" is paramount in Borges's "Our Poor Individualism," "From Somebody to Nobody," and "New Refutation of Time." But perhaps the theme is nowhere more evident than in "Everything and Nothing" (*L*, 248–49). A young man went to London at the age of twenty-odd years. He had already "become proficient in the

habit of simulating that he was someone, so that others would not discover his condition as no one" (*L*, 248). In London he became an actor and could now play at being another person before spectators who played at perceiving him to be that other person. After a long, illustrious career, and at the point of death, he came into the presence of God and petitioned Him that he might finally become someone: "The voice of the Lord answered from a whirlwind: 'Neither am I anyone; I have dreamt the world as you dreamt your work, my Shakespeare, and among the forms in my dream you are, who like myself are many and no one' " (*L*, 249).

This Schopenhauerian notion of the one being all and the all one—or none—recalls, of course, much Oriental thought. Surprisingly, and as I shall illustrate further in Chapter Six, it is also a concept recently suggested by some contemporary physicists. Of note is the remarkable passage by Erwin Schrödinger (1967, 138):

> The reason why our sentient, percipient and thinking ego is met nowhere within our scientific world picture can easily be indicated in seven words: because it is itself that world picture. It is identical with the whole and therefore cannot be contained in it as a part of it. But, of course, here we knock against the arithmetical paradox; there appears to be a great multitude of these conscious egos, the world however is only one. This comes from the fashion in which the world-concept produces itself. The several domains of "private" consciousness partly overlap. The region common to all where they all overlap, is the construct of the "real world around us."

Is this not Cantor's paradoxical sets in a new garment? How can there be a region common to all conscious egos? How can consciousness be one and at the same time many? The very notion borders on the outlandish, yet the suggestion is put forth in all seriousness.

An intriguing counterpart to this theme, the relevance of which will become evident as this discussion continues, is found in Borges's "Funes the Memorious" (*L*, 59–67). Funes is capable of seeing only particulars. During the early stage of his life, he lived like all men, in a dream: looking without seeing, listening without hearing, forgetting almost everything. After having been one fateful day thrown by a horse, he discovered that his perception and memory had become infallible. He could at a glance take in all the leaves, branches, contours on the trunk,

etc., of a tree, and years later recall them to memory perfectly. The problem was that his memory became a garbage heap. It contained an indefinite number of individuals, but he was incapable of "ideas of a general, Platonic sort." It seemed strange to him that a dog seen at 3:14 P.M. from the side was considered to be the same dog seen at 3:15 P.M. from the front. Conceiving number as an ordered series was for him impossible. He simply memorized each number without establishing the necessary relations between them. In fact, he once developed his own alternative number system consisting of arbitrary names in place of every number, which for him was just as effective. Funes, in short, was unable to think, for to think "is to forget differences, generalize, make abstractions. In the teeming world of Funes, there were only details almost immediate in their presence" (*L,* 66).

Funes, it appears, is not even capable of higher animal forms of selection. He either sees all or nothing at all; he remembers aggregates of particulars without being able to isolate any of them. He is, in other words, the consummate nominalist, a superempiricist. A hypothesis, theory, conjecture, even a beginning, would be for him impossible. Before there can be anything at all, even before there can be no-thing, there must be something, and this some-thing must be a selection, an abstraction from the whole. If we were absolutely pure empiricists, in essence we would be less than human; we would be like Funes (or perhaps we would be more than human; we would be like Dunne's serial observer, or like God). Although our imaginary Funes escaped the human penchant for hypothetical abstracting, we obviously cannot. This is actually fortunate for us. One might conceive Funes's teeming jungle to be perpetually in motion, a myriad array of differences. But surprisingly, Funes, ultimately and unlike us, is indifferent toward his world:

> Then I saw the face belonging to the voice that had spoken all night long. Ireneo [Funes] was nineteen years old; he had been born in 1868; he seemed to me as monumental as bronze, more ancient than Egypt, older than the prophecies and the pyramids. I thought that each of my words (that each of my movements) would persist in his implacable memory; I was benumbed by the fear of multiplying useless gestures. (*L,* 66)

Funes represents a closing of the circle. His unlimited perception and memory of the world is total. Everything is *actually*

there. By the same token, since nothing is/can be abstracted, everything is reduced to essentially the same level. Funes's perception, then, approaches total indifference, or perhaps it *is* total indifference. That is, as differences become increasingly finer, finally it becomes much more difficult for one thing to stand out (be abstracted) against the background of everything else.

However there is a paradox inherent in Funes's apparent dilemma, which Borges does not reveal. If Funes can, in simultaneity, perceive every minute particular of a tree in one instant and commit it to memory, he must subsequently be able to perceive it again to be a different (or slightly variant) tree, in all its details, for it has suffered an ever-so-small change. So he undertakes another peceptual grasp, and then another, and another, and at each "instant" he sees a slightly different collection of particulars before him. But the question is, How is it possible for him to detect any movement at all? In other words, if he sees everything all at once as an aggregate of particulars, and if he cannot abstract anything, then he is incapable of seeing one particular against the background of the whole, and hence he cannot detect a change in that particular while holding the whole in check as an unchanging entity. What he does perceive, and the only thing he can perceive, is, so to speak, a succession of static "slices," a series rather than a continuum. But in this event there can be for him no change, no time in the conventional sense, the same dilemma presented by Zeno's paradox of movement, for which Borges has a special affinity and which has resisted complete resolution over the centuries.[19]

Now, if we translate Funes's dilemma to Borges's essay on Dunne, it can be conjectured that if Dunne's infinite regress of consciousness represents a series of discrete terms (i.e., consciousness$_1$, consciousness$_2$, consciousness$_3$, . . . consciousness$_n$), then it is equivalent to Funes's incapacity to step outside a particular perceptual grasp, and Zeno's arrow paradox prevails—hence Dunne's consciousness is paralyzed. On the other hand, if the infinite regress of consciousness is a continuum, then all consciousness overlaps to become one, as Schrödinger posits and as Borges often implies.[20] In this sense, ideally, Funes is the equivalent of a superhuman observer, the composite of all possible observers, for he is capable of seeing everything in simultaneity. The problem is that, rather than his perception representing a unifying whole, there is no more than a disrupted series.

The Demise of Totalizing Quests

Very significantly, in this respect, the narrator of "The Zahir" admits at the outset that: "I am no longer the 'I' of that episode; but it is still possible for me to remember what happened, perhaps even to tell it. I am still, however incompletely, Borges" (*L,* 156). Incompletely he is Borges, for the "I" has become lost in the infinitesimal interstices, it has become incapable of knowing the world *apart from* individuals, *apart from* The Individual: the Zahir. We are later told that images other than the Zahir come to the narrator fragmentarily, as if from far away. Tennyson is then evoked, who once said that if we could understand a single flower, we would understand the world and ourselves. Perhaps, the narrator speculates, he meant that there is no single and even insignificant fact "that does not involve universal history and the infinite concatenation of cause and effect" (*L,* 163). Schopenhauer is also mentioned, as are the Kabbalists, who pretend that each man is a microcosm, "a symbolic mirror of the universe" (*L,* 163). Finally, by 1948, the narrator declares, he will be an invalid, he will not know who Borges is. He will not perceive the universe, only the Zahir, passing from a "highly complex dream" (reality) to a dream of "utter simplicity." The Zahir, to reiterate, affords a series of particular perspectives—like a series of snapshots (or, at various levels, Funes's perceptual grasps, Dunne's serial consciousness, and dream images). But it is incapable of illustrating the unity of all things. In contrast, Schrödinger's consciousness, like the Aleph, continues to nurture one of our vain desires for our own self-fulfillment, ironically.

In sum, Hayles demonstrates that Borges is attracted to Cantor's work because he saw in it the possibility for creating new kinds of paradoxes, recursive "strange loops," wherein everything is connected to everything else. Borges is also fascinated by Gödel-like paradoxical systems, especially in his Library. Dunne and other writers of the bizarre, such as Bradley and Hinton, obviously pique Borges's interest for similar reasons. But perhaps Borges is drawn to these authors because he, unwittingly, intuits more than he reveals. Perhaps they exercise a strange power over him because he senses an affinity between himself and their curious pretenses. I refer to his own déjà vu experience found in *The History of Eternity* and "New Refutation of Time," which is, itself, of Cantorian vintage and intimately related to the Aleph metaphor. One evening in Barracas, Borges, having nothing to do, went out after dinner for a walk. He had no set destination, and followed a random course,

but a familiar gravitation began drawing him toward certain sections, and finally to a street corner, where after some breaths of fresh air and contemplation of the surroundings, he realized that it was the scene of some thirty years ago. He was in the nineteenth century: the phrase was more than mere words, it had emerged into reality:

> I felt dead—that I was an abstract perceiver of the world; I felt an undefined fear inbued with knowledge, the supreme clarity of metaphysics. No, I did not believe I had traveled across the presumptive waters of time; rather I suspected I was the possessor of the reticent or absent meaning of the word *eternity*. Only later was I able to define that imaging. (*OI*, 180)

Borges wisely realizes the fallacy of Dunne's argument, though he would like to be a believer, for the thesis is so splendid that any error it contains is, Borges says, "magnificent." Dunne's thesis promises the logically impossible, which, nonetheless, has endured as the focus of humankind's feeble hopes over the centuries, a hope intermittently kindled by various forms of mysticism. Yet Borges has denied having any mystical experience, except, he qualifies, perhaps in "feeling in death" (Christ, 1972, 405). And yet, he never ceased nurturing a hope for the impossible grasp of the Totality. It is safe to say that few have struggled to understand the universe's random babble more fervently than Borges. Any lack of success does not reduce the merits of his efforts, but heightens them, for meaning exists essentially in the struggle.[21]

Interlude

Upon reading any one of the numerous popularizations on the "new physics," one cannot help but marvel at the apparently absurd, mind-boggling interpretations of what one ordinarily conceives to be one's everyday world of trees, cars, pavement, TV sets, filthy air, bookstores, football stadiums, and empty beer cans. These bizarre descriptions of "reality" have been in part responsible for recent interest in scientific creativity, the logic of scientific "discovery," and empirical studies on how mathematicians and scientists think. In this brief interlude, I summarize, in general terms, the emergence of the "new physics." Then, with the creative process in mind, I focus on both the pure scientist and Borges—both supreme fiction-makers—as aesthetic and epistemological chameleons, the demands of whose craft often compels them either to exist in two or more "realities," or to become ephemerally suspended in between. This will serve as a preparation for the chapters that follow on the universe of Einsteinian physics and quantum mechanics and their relevance to key Borges works.

Interlude

Mathematical physics translates the saying of Heraclitus, "All things flow," into its own language. It then becomes, All things are vectors. Mathematical physics also accepts the atomistic doctrine of Democritus. It translates it into the phrase, All flow of energy obeys "quantum" conditions.
—Alfred North Whitehead

Number Power/ Word Power

1 Gene H. Bell-Villada (1981, 266) remarks that Borges's bleak philosophical nihilism by no means enjoys universal acceptance. . . . [S]uch views are hardly conceivable and would be less than absolute from the pen of a practicing politician or research scientist, both of whom accept a body of knowledge and its uses. The fact looms large that Borges, his celebrated erudition notwithstanding, is scarcely informed about natural science—its mode of knowing, its praxis—while his sense of twentieth-century social thought is particularly limited.

The problem with Bell-Villada's view is that most contemporary scientists—and I speak of pure scientists here—no longer nurture categorical beliefs in absolutes; their science is much like that predicted by Nietzsche almost a century ago. And while it is true that Borges professes to know little about natural science, as we shall note, he nonetheless shares much of the spirit of twentieth-century physics.

Before the new scientific worldview could begin to emerge, disintegration of the classical model was to be realized. Thermodynamics began eroding the mechanistic Laplacean ideal of absolute knowledge about the mid-nineteenth century. Toward the end of that century, the first indication of a breakdown in the mechanical paradigm followed investigations in magnetism and electricity by Michael Faraday, James Clerk Maxwell, and Heinrich Hertz. In fact, Maxwell's equations so refuted the foundations of Newtonian physics that at the outset many leading physicists rejected them outright. Only after decades were his radical ideas fully assimilated by the scientific community. Although Maxwell served as a battering ram to dislodge chinks from the wall of classical mechanics, he alone was not sufficient. Many still considered that the foundations of classical physics remained unchallenged; consequently, whenever an explanation

could not be derived from empirical observation, one was conveniently invented. The most noteworthy of such inventions, in a manner of speaking equally as absurd as the "phlogiston" of old, was the classical "ether" hypothesis. The "ether" was devised for the planets to "swim in," to "constitute electric atmosphere and magnetic effluvia, to convey sensations from one part of our body to another, till all space was filled several times over with ether" (Maxwell, quoted in Jeans 1930, 87).

Interlude

It was not until after the publication of Einstein's special theory of relativity in 1905 that the monster, "ether," was finally banished from the kingdom. But at what price? Einstein, with Max Planck, was instrumental in laying the cornerstone for quantum theory, far more revolutionary than relativity theory, which finally spelled death to the smug absolutes of classical physics, perhaps for all time—and much to the chagrin of Einstein himself, for he never gave up his quest for certainty. In Einstein's own words (1949a, 45), "It was as if the ground had been pulled out from under one, with no firm foundation to be seen anywhere upon which one could have built." After the full impact of relativity, and especially quantum theory, had been felt, there was little left of the belief in the classical image of an objective reality as adequately registered by the senses. It was not long before Eddington (1958a, 57), not unlike other scientists, arrived at the conclusion that the so-called "laws of nature" correspond to a priori considerations and conceptions of "reality," and hence they are by and large subjective. The can of worms was finally opened, and there were more than a few apprehensions about this new state of affairs. After a couple of decades, and especially between 1924 and 1929, when Niels Bohr's complementarity and Heisenberg's uncertainty principle were formulated, it seemed that well-known ancient formulae such as Heraclitus's "all things flow," Zeno's paradoxes, and the Pythagorean notion of structure as music were beginning to appear less like sophistry and more like keen insight and valid descriptions of mental experiences.

A difficulty in presenting the emerging scientific view stems from that ancestral humanistic notion of the scientist—which has unfortunately survived in many circles—as a person of nothing but facts and hard-nosed calculation, empty of all poetry and sense of wonder, whose discovery process is akin to a computer search and whose arrival at truth is by means of a mechanical procedure. Such were the views of Blake, Keats, and Shelley, but not, interestingly enough, of Coleridge. In

Interlude order to dispel this myth of the scientist, one need look no further than to what is commonly termed the scientist's "method." It has been tacitly assumed by some philosophers of science that the idea of a fixed scientific method is no more than myth—Francis Bacon to the contrary. For example, Alexandre Koyré (1939) argues convincingly that Galileo, supposedly the father of modern science, neither dropped stones from the tower of Pisa nor used inclined planes. Feyerabend (1975, chaps. 6–12) concurs, demonstrating further that Galileo availed himself of ingenious thought experiments, of his telescope, which altered the core of sensory experience, of ad hoc explanations, and even puzzling and unexplained phenomena. Indeed, theoretical science is not an algorithmic game, for "there is no philosophical highroad in science, with epistemological signposts. No, we are in a jungle and find our way by trial and error, building our road *behind* us as we proceed. We do not *find* signposts at crossroads, but our own scouts *erect* them, to help the rest" (Born 1943, 44).

Methodological imperatives will only serve to limit the pathfinding enterprise of the scouts. Sense-data, observations, and faulty pointer readings are sometimes conveniently overlooked; facts are on occasion ignored. And initially fuzzy concepts often prove superior to rigorously constructed theories. Feyerabend (1975, 24n) tells how, for example, Bohr would invariably begin with an apparent paradox, mulling it over in his mind, gradually teasing a few answers from it, to produce a solution eventually. Great scientists do not follow a predetermined method, but whatever activity brings them success. In Einstein's (1949b, 684) words, the scientist

> must appear to the systematic epistemologist as a type of unscrupulous opportunist: he appears as *realist* insofar as he seeks to describe a world independent of the acts of perception; as *idealist* insofar as he looks upon the concepts and theories as the free inventions of the human spirit (not logically derivable from what is empirically given); as *positivist* insofar as he considers his concepts and theories justified *only* to the extent to which they furnish a logical representation of relations among sensory experiences. He may even appear as *Platonist* or *Pythagorean* insofar as he considers the view-point of logical simplicity as an indispensable and effective tool of his research.

Surprising as it may seem, is this not Borges's modus operandi as now a nominalist, now a realist, usually an idealist but

sometimes not quite, never a positivist though at times manifesting vague trappings of positivism? If Borges uses mathematical constructs and metaphysical doctrines primarily to satisfy his aesthetic purposes, the theoretical scientist uses them to get her own results. Ultimately the scientist's "facts" are inseparable from her metaphysical ideas, and Borges's metaphysics is intimately linked to the shape of his essays, prose, and poetry. Perhaps, in the final analysis, Borges's fictions, or any other fictions for that matter, are in some respects not necessarily any less "real" than the scientist's theories and models, derived from her freely wandering mind rather than directly from the "facts" before her. Once again, Einstein (1934, 30) offers a provocative statement: "If you want to find out anything from the theoretical physicists about the methods they use, I advise you to stick closely to one principle: don't listen to their words, fix your attention to their deeds." These deeds tend to indicate that the theoretical scientist is a very imaginative person who would like to think her constructs correspond to "reality"; that is, to the everyday world in which she functions. Nevertheless, at the outset her theories and models are purely imaginative acts.

But this is only one perspective of the Janus-faced scientist. First and foremost, she must possess the capacity for *wonder* at the very fact that the universe exists at all, and this sense of wonder inevitably leads to the construction of problem situations, the solutions to which help explain the universe. In contrast, most people live and die without giving the most fundamental problems a moment's thought. Borges is keenly aware of this, responding thusly to Burgin's (1968, 6) inquiry about popular conceptions of time, space, and infinity:

> Well, because they [the general public] take the universe for granted. . . . They take themselves for granted. That's true. They never wonder at anything, no? They don't think it's strange that they should be living. I remember the first time I felt that was when my father said to me, "What a queer thing," he said, "that I should be living, as they say, behind my eyes, inside my head, I wonder if that makes sense?" And then, it was the first time I felt that, and then instantly I pounced upon that because I knew what he was saying.

The scientist's problem situation derived from her sense of wonder and her questioning of nature, however, does not lead to a solution necessarily corresponding to "reality." The wonder

Interlude is in its very inception basically mental, not the exclusive product of sensory experience. As Karl Popper (1972) argues, the problem situation is conceived "hypothetico-deductively." And the physical concept generated from it is a "free creation of the human mind" rather than being "uniquely determined by the external world" (Jammer 1962, 4; see also Einstein, 1949a, 7).

In addition to possessing the sense of wonder, the scientist, like the artist, has always been attracted to beauty, harmony, and simplicity, even in this century of uncertainty. According to P. A. C. Dirac's (1963, 47) maxim, "It is more important to have beauty in one's equations than to have them fit experiment." Although not all physicists, I have been told, declare such undying faith in the "beauty of one's equations," they all tend in that direction. For instance, Heisenberg (1972, 68–69) remarks, in rather circular fashion, "If nature leads us to mathematical forms of great simplicity and beauty . . . we cannot help thinking they are true, that they reveal a genuine feature of nature." More recently, John Wheeler has reiterated the same sentiment: "The beauty in the laws of physics is the fantastic simplicity that they have. . . . [W]hat is the ultimate mathematical machinery behind it all? That's surely the most beautiful of it all" (Buckley and Peat 1971, 60). Einstein (1949a, 33) lists three elements of scientific beauty in a single formula: "A theory is the more impressive the greater simplicity of its premises is, the more different kinds of things it relates, and the more extended is its area of applicability." Finally, the mathematician-physicist Henri Poincaré (1958, 8) divulges the scientific motives for his quest: "The scientist does not study nature because it is useful to do so. He studies it because he takes pleasure in it; and he takes pleasure in it because it is beautiful. If nature were not beautiful, it would not be worth knowing and life would not be worth living."

Arthur Koestler (1964, 246) goes so far as to propose that science manifests a tendency to become fashionable, like the arts. There is in all probability a degree of truth to this notion, but the fact remains that at the leading edge, both the sciences and the arts depart violently from fads and fashions. It is this leading edge we must be concerned with, for scientific theories, as free creations of the mind, are "idealizations which most likely become less applicable to reality as they become more complete, and hence, contrary to Descartes, nothing is more misleading than a clear and distinct idea, for if it is supposedly clear and distinct, it has no necessary bearing on reality, and if it

supposedly has a bearing on reality, it is invariably subject to alterations" (de Broglie 1954, 219).

Supposedly "clear and distinct ideas" must be perpetually subject to alterations, since, given the myriad array of scientific explanations, past, present, and future, each and every one is at least partly false—except, of course, the theories that in a given moment in history are considered to be "true." In other words, there exists, at a given moment, potentially an infinity or at least a quasi-infinity of possible theories from which to choose. And since each and every theory is, from one of a vast number of perspectives, false, then it is impossible to know if an explanation corresponds to "reality" or not. Borges (*OI,* 137), in this vein, says with respect to biography writing:

> Reality is so complex, history is so fragmentary and so simplified, that an omniscient observer could write an indefinite, and almost infinite, number of biographies of a man, each of which would emphasize different facts; we would have to read many of them before we realized that the protagonist was the same man. Simplify a life overmuch: imagine that thirteen thousand facts describe it completely. One of the hypothetical biographies would record the series 11, 22, 33, . . . ; another, the series 9, 13, 17, 21 . . . ; another, the series 3, 12, 21, 30, 39 . . .

It becomes evident that each "hypothetical biography" is a choice, a selection, an abstraction. Each is partly arbitrary, partly determined by the biographer's mind-set and his cultural milieu. At best it is a dismally incomplete picture.

Regarding contemporary physics, the problem of abstraction is even further aggravated, for at the interface between the scientist and nature lies, precisely, mathematics—mental constructs par excellence. Recent development of physics has required, for its theoretical formulation

> a mathematics that gets continually more advanced. This is only natural and to be expected. What, however, was not expected by the scientific workers of the last century was the particular form that the line of advancement of the mathematicians would take, namely, it was expected that the mathematics would get more and more complicated, but would rest on a permanent basis of axioms and definitions, while actually the modern physical developments have required a mathematics that continually shifts its foundations and gets more abstract." (Dirac, quoted in Bellone 1980, 22–23)

Interlude

While this growing power of theory (mental constructs) and consequently of mathematics over the world has been of concern to some, others concur with Dirac that it is, nevertheless, "more important to have beauty in one's equations than to have them fit experiment." Certainly, in this light, one can conjecture that the physicist abuses his right to mathematics, and this is most likely to a degree correct. When the physicist discovers a connection between observed results that resembles a connection well known in mathematics, he often tends to "jump to the conclusion that the connection *is* that discussed in mathematics simply because he does not know of any other connection" (Wigner 1969, 131–32). Perhaps the physicist is by nature an overaggressive "opportunist," to use Einstein's term; or perhaps he is at times merely an "irresponsible person," according to Eugene Wigner (1969, 132). Yet the fact remains that his mathematical formulations lead "in an uncanny number of cases to an amazingly accurate description of a large class of phenomena" (Wigner 1969, 132).

The question might now be, Can a mathematical description be truly accurate, or is it no more than an apparently reliable report *from a particular perspective that would be refuted by a contradictory mathematical formulation from one of an infinity of other possible perspectives?* The answer must lean toward the second alternative, which brings us back to Borges's commentary on the infinite variety of biographies written by an omniscient observer, and in general, to Borges's notion of fictionality. Significantly, in a somewhat Borgesian vein, Wigner (1969, 139) modestly admits, with respect to the use of mathematics, that

> [t]he miracle of the appropriateness of the language of mathematics for the formulation of the laws of physics is a wonderful gift which we neither understand nor deserve. We should be grateful for it and hope that it will remain valid in future research and that it will extend, for better or for worse, to our pleasure even though perhaps also to our bafflement, to wide branches of learning.

Let us paraphrase this statement in order to render it notably commensurate with Borges's thinking:

> *The miracle of the appropriateness of natural language for the formulation of fictions is a wonderful gift which I [Borges] neither understand nor deserve. I should be grateful for it and hope that it will remain valid in future writing and that it will extend, for better*

or for worse, to my pleasure even though perhaps also to my bafflement, to increasingly broader conceptions of fictionality. **Interlude**

Whether speaking of discourse and mathematical constructs hopefully corresponding to the world, or of fictions, the same conclusion inheres.[1] Our capacity for language and mathematics has liberated us from the mundane constraints of time and space. It has also solidly locked us into our linear mode of thought, which is becoming increasingly akin to "computer thought," thus rendering inaccessible the regenerative power of number and the word—dear to the ancients—which precipitates matter from pure form. We have gained in our ability to transform the world, whether for good or for bad, but we have lost a certain power of resolution: though we "hyperfictionalize" to the extreme, we realize we can no longer believe with the assurances of the past. I now turn to a characteristic of these "hyperfictions" that ties scientific activity to literary activity.

But other men are oblivious of what they do awake, just as they are forgetful of what they do asleep.
—Heraclitus

[2] Borges (Burgin 1968, 156) believes that philosophy, over the long haul, gives the world a kind of "haziness," which is actually preferable to absolute clarity—recall Benardete's "sludge." "Haziness" serves to dissolve "reality," and since "reality" is not always pleasant, we can be aided by this dissolution. Borges, it seems, hangs between "reality" as a dream, an "insomnious dream" so to speak, and "reality" as empirical, between metaphysics as "truth" and as fiction. Indeed, what is gradually being revealed here is that the contemporary physical scientist, like Borges, is irradicably "schizophrenic." In fact, we all are, with respect to the twentieth-century scientific cosmology vis-à-vis our everyday working, eating, playing, and sleeping world. Were the scientist hoping to produce not only a mathematical model of a worldview but also a worldview that could be, as it were, commensurate with our everyday commonsense world, he would be treading a path toward frustration. Einsteinian space-time and especially quantum theory as yet have had little impact on our commonsensical notion of things: as such they are paragons of "hyperfictionalization." Of course,

The World as Dream

Interlude one might retort that it is too early to pass judgment, since a couple of centuries transpired before the Copernican heliocentric cosmology became "commonsensical." Perhaps. But the fact remains that the space-time of relativity and quantum theory are, as we shall observe, not only unvisualizable but virtually inexpressible in natural language; they appear to defy all logic and reason, so divorced are they from our everyday notions and sensory experience. The scientist may describe as best he can one of the current hypotheses to a layperson and end up with, "Isn't that fantastic!" to which the layperson responds, "Yes, it's intensely interesting," and both felicitously go home to dinner, having forgotten about the whole affair.

Koestler (1963) calls the scientist a "sleepwalker," a semi-awake person who tinkers, dabbles, and plays. This is comparable to what I termed an "insomnious dreamer." Somewhat like a child in a game of make-believe, or like an accomplished fabulist—and I do not write this in a derogatory sense—the scientist often tries things out simply to see if they work, and finally, if she is fortunate, she hits upon an answer. But she could not see where she was going, nor could she know what she was doing, except in retrospect. She exists, in other words, at two levels: the person of day-to-day living, and the other, the person of her mental constructs, who maintains constant vigil over herself as if in a dream.

Here, the reader familiar with Borges will recognize "Borges and I" (*L*, 246–47), which is characteristic, to a greater or lesser degree, of all fiction-makers. The "other one, the one called Borges," is the one things happen to. The everyday Borges, on the other hand, is the one who walks the streets of Buenos Aires, stops for a moment to gaze at various scenes, the arch of an entrance hall, or perhaps the grillwork on a gate. He knows of the "other Borges" only by the mail he receives, his appearance on a list of professors, or in a biographical dictionary. The flesh and blood Borges realizes that he is "giving over everything to the 'other,' though I am quite aware of his perverse custom of falsifying and magnifying things" (*L*, 246). Borges recognizes himself less in the "other's" books than, say, in the "laborious strumming of a guitar" (*L*, 246). Years ago he endeavored to free himself from his "other" by writing stories of games, time, and infinity, but now all this belongs to the "other." His life is thus a flight, he is losing everything to oblivion, or perhaps to the "other." And the rumination ends: "I do not know which of us has written this page" (*L*, 247).

This beautiful counterpart to the scientist as "sleepwalker" or dreamer, gives insight into the human brain's remarkable capacity for ordering the disordered, producing structure out of chaos. Even during sleep, the quest for harmony persists. The nightly drama may seem to be random and chaotic, totally out of control, but, examined in fuller detail, patterns inevitably emerge, and the complex of dream images is interrelated. It has been suggested that dreams reveal some of the basic characteristics of a theory-making machine; or, by extrapolation, a fiction-making device (Campbell 1982, 230–37). Images are fitted to images in an overall pattern of relations, and new patterns are generated, apparently from nowhere. They are grasped as if "bolts from the blue," free flights of the imagination. Borges, significantly, tells us that literature is none other than "controlled dreaming." Our past is "nothing but a sequence of dreams. What difference can there be between dreaming and remembering the past? Books are the great memory of all centuries" (Alifano 1984, 34).[2]

Regarding this same theme, in "The Other" (*BS*, 11–20) the young teenaged Borges mysteriously appears from nowhere and sits down with the aged Borges on a bench facing the Charles River in Cambridge in February 1969. After they become properly acquainted, the younger Borges speculates that he is most likely dreaming his older counterpart. The latter objects: "If this morning and this meeting are dreams each of us has to believe that he is the dreamer. Perhaps we have stopped dreaming, perhaps not. Our obvious duty, meanwhile, is to accept the dream just as we accept the world and being born and seeing and breathing" (*BS*, 13). He then reveals that his dream has lasted some seventy years. After a brief conversation the younger fellow takes his leave, never to show up again. The elder Borges broods over this for days, finally to discover, he believes, the key. The meeting was "real." But the young man was dreaming, while the aged Borges was not, and this "explains how he was able to forget me; I conversed with him while awake, and the memory of it still disturbs me" (*BS*, 20). Forgetfulness is notorious where dreams are concerned; memory favors wakefulness. The two men represent incompatible states of mind in this regard. Moreover, they apparently could not adequately understand one another: "We were unable to take each other in, which makes conversation difficult. Each of us was a caricature copy of the other" (*BS*, 18). A half a century separates them. The young Borges of Borges's memory is now vastly different from the young Borges as he was,

Interlude

Interlude and the former is incapable of comprehending the latter. The important point is that if, metaphorically speaking, one is dreaming and the other awake, are the aged Borges's memories in essence any different from a dream? If not, then it is *as if* he were dreaming also, and if so, then the two did not exist in incommensurable frameworks.

Borges's metaphor bears on the age-old problem of knowing whether or not we are dreaming.[3] Briefly, the statement "I am dreaming" is either patently false or nonsensical. If I utter the statement while awake, as supposedly was the elder Borges, my statement is obviously false. On the other hand, if I am asleep, as we assume the younger Borges to be, I cannot make the statement *about* my dream state *as if* I were outside that state, hence "I am dreaming" would be in this regard a nonsensical statement. Suppose, then, that I am asleep and somehow dreaming *about* myself dreaming. I might say "I am dreaming" in my sleep, as it were, but if so, this utterance is *about* a dream state *as if* from a wakeful state, so it is also nonsensical from a literal point of view. What, in this light, is meant by Borges's notion of "controlled dream" or "hallucination"? If I am actually dreaming, then obviously I cannot be in control of my dream, for control exists solely from the waking state. But this explanation is materialistic, even behavioristic, and therefore far removed from Borges, who says that according to idealism, "the words 'live' and 'dream' are rigorously synonymous" (*L,* 164). For Borges, "controlled dream" is not a contradiction or an oxymoron but more appropriately, and when viewed from an all-encompassing framework, *a strange sort of tautology, for dream state and wakeful state are not absolutely incommensurable.*

To dream, in the sense of "controlled dreaming," is to idealize, and ultimately, if the dream is fictionalized, to turn the world *into* words, to assume the world to be nothing *but* words. When one says all that can be said about the world, this concoction of words *is/becomes* the world—but not in the above sense of "word magic"; it is simply that words are virtually all we have to go by. Language abstracts from the world. To use words in a certain way is the equivalent of forgetting what could otherwise have been said (abstracted). And since dreams are more forgetful than wakefulness, to equate thinking with dreaming and thus abolish the distinction between the "real" and the hallucinatory seems to deflate the world quite violently. But the concomitant inflation of fiction is only temporary precisely because it is fictive.

Berkeley, in a certain stretch of the imagination, believed our fictions *are* our world; they are the simultaneity of the

thoughts-signs of God, who is the ultimate dreamer. Berkeley's curious metaphysics has been generally unpalatable to all but a few hardy idealists. James Jeans, however, was one of the first contemporary physicists indirectly to shed light on idealist philosophy when he declared:

> Today there is a wide measure of agreement which on the physical side of science approaches almost to unanimity, that the stream of knowledge is heading towards a non-mechanical reality; the universe begins to look more like a great thought than like a great machine. Mind no longer appears as an accidental intruder into the realm of matter; we are beginning to suspect that we ought rather to hail it as the creator and governer of the realm of matter. (Jeans 1930, 158)

Jeans went on to argue that since the universe has the appearance of a "Great Thought," thought must exist in the mind of a "Great Mathematician." Repugnance over the spiritual implications of this notion alienated many scientists during the early days of relativity and quantum theory, but today it appears to bear more than a slight degree of truth. Indeed, the power of mathematics has been outstanding, perhaps the supreme example being $E = MC^2$, an elegantly simple formula, which, when set apart from a background of dizzyingly intricate symbols, so controls "reality" as to render humankind virtually helpless. Perhaps, as Einstein has mused on numerous occasions, the stately dream of the ancient Pythagoreans may yet be realized. The Pythagoreans' dream *was* a dream, a "controlled hallucination" or "hyperfiction," that they desired to interject into the world; that is, they attempted to beat "reality" into shape so that it might conform to the dream. Are the harmonious spheres or the labyrinthine confusion of symbols in today's physics a pattern that actually mirrors the world? To these questions, forthcoming answers are more than can be expected. Yet cognizance of the interactive play between desire and the "real," dream and the world, coupled with our feeble efforts to understand and the recalcitrant opposition of that which must be understood, at least serve to prevent the quest from dying altogether.

Borges often elevates the reader to an awareness of this interactive play between dream and what is ordinarily construed to be the "real," especially in "The Circular Ruins," and in another story, "The Waiting" (*L*, 155–68). According to the latter tale, a man calling himself Villari—because the name disturbed

Interlude him—checks into a small room, from which he does not depart during the first few weeks. He is pursued by a nightmare, which he hopes to dispel. Years of solitude have demonstrated to him that in one's memory timelessness prevails; all days tend to be the same, yet each day inexorably brings surprises. Actually, he expects an unexpected surprise: the realization of his dream, when he will be killed by two assassins. Heraclitus once counseled that unless one expects the unexpected one will never find truth. Truth, however, for Villari, cannot be so clear-cut. His dreams generate his expectations, which are interpolated into his world, which in turn threatens to become his dreams, because Villari speculates that he "*had already died* and in that case this life was a dream" (*L,* 167). He is disturbed by this thought, for he cannot decide "whether it seems a relief or a misfortune" (*L,* 167). He finally decides the question is absurd and discards it. He then endeavors to live in the simple present, with no memories or anticipation, like the dog he has befriended. At times, in his weariness, he feels content; but this feeling is only ephemeral. Finally, one morning, he is awakened by the presence of two men. Alejandro Villari and a stranger have found him at last. He asks them to wait and turns to the wall as if to continue sleeping; then he is shot. Villari, the dreamer, is finally confronted by Alejandro Villari, his double, the dreamt. The narrator comments on the protagonist's final but futile act:

> Did he do it to arouse the pity of those who killed him, or because it is less difficult to endure a frightful happening than to imagine it and endlessly await it, or—and this is perhaps most likely—so that the murderers would be a dream, as they had already been so many times, in the same place, at the same hour?
>
> He was in this act of magic when the blast obliterated him. (*L,* 168)

Villari's nightmare had been "reality" from the very beginning, according to the gaze of an omniscient being.

Of course, the theme of dream within "reality" and "reality" simultaneously within dream, both of them making up a totality, corresponds nicely with Cantor sets as discussed in Chapter Three. The important issue here, however, is another. Fiction, that is, "controlled dream," mediates between what the physical world is, Kant's *Ding an sich,* which we are destined never to know, and pure, unfettered idealism, which Borges cannot accept *in toto.* Villari, for instance, often attended the mov-

ies. He would witness tragic stories of the underworld, some containing the images of his former life, but he "took no notice of them because the idea of a coincidence between art and reality was alien to him. . . . Unlike people who read novels, he never saw himself as a character in a work of art" (*L,* 166). If Villari disavowed the mimetic notion of art, he also rejected the conjunction of dream and art, since he could not bring himself to identify with the artwork. For Villari, dream is disjoined from "reality." Or at least he desired to make it so in order to escape the "reality" of his nightmares. Villari is the inverse of Hladik, who, in a manner of speaking, expected the unexpected moment and therefore was not excessively surprised to be surprised, for he (by the grace of God) was perfectly in control of his "dream." Villari, in contrast, hopelessly attempted to divorce both art and dream from "reality," but in the end his greatest fear was confirmed: dream, art, and "reality" are ultimately one.

It can be said that the contemporary physicist, especially following Jeans's comment, is a literary and metaphysical spirit subtly disguised as a mathematical spirit. Conversely, Borges, as some critics tend to note, is a mathematical spirit disguising itself as a literary spirit. Both efforts revolve around the axis of mathematics. Mathematics, unlike Villari's dream, entails symbols of condensation, or "metaphors": controlled "hyperfictions" that do not correspond to the "real," but which can be, and invariably have been in modern Western thought, used to pattern the "real." In regard to such symbols, let us return to Carlos Argentino Daneri of "The Aleph" for a moment. Daneri has written an infinite, or at least a quasi-infinite, poem, which supposedly gives account of the entire world. He does not make use of symbols of condensation that are capable of encompassing broad conceptual and emotive wholes. His poetry, in Cantorian fashion, consists of putting a series of words into one-to-one correspondence with another set: the furniture of the world. This can ideally be accomplished, for if the world is an infinite set, and language—or Daneri's poem—one of its subsets, then the subset is also infinite. This ideal possibility changes the role of artistic choice in literary composition, for if literature could by any stretch of the imagination be nothing more than a string of words mimetically placed in correspondence with the world, choice would become severely curtailed, and symbols of condensation would be prohibited.

And choice is an important issue here. In her study, Hayles (1984, 158) refers to the "axiom of choice" in set theory,

Interlude which has been of debatable value. This axiom consists of the arbitrary selection of an element from each of a given number of sets and subsequent use of this combination of elements to produce a new set. Hayles then demonstrates the relevance of the axiom of choice to the conflict between the narrator's notion of the ideal poem and that of Daneri. For the narrator of "El Aleph," in contrast to Daneri, a series of one-to-one correspondences is first refuted as a valid descriptive mechanism, then it is reconstituted by means of a single condensed element chosen by the poet. Rather than time-bound linearity, atemporal synchronicity is the goal; rather than a series, a singular symbol of condensation, the Aleph, which simultaneously signifies the all, is posited. But since the all, in Borges's conception of things, cannot be available in one simultaneous grasp, a chosen symbol can never be coterminous with the "real."

Symbols of condensation have invariably been selected for use in the mathematician's and especially the physicist's playgrounds. I am not speaking here of general cosmological models equivalent to the Aleph (i.e., Jeans's universe as a Great Thought, the Newtonian-Cartesian machine model), but of relatively modest symbols that pattern microdomains. Consider William Rowe Hamilton's "quaternions," for instance. After much work, Hamilton conjectured that in three-dimensional space a vector might be represented by a set of three numbers, a triplet, just as a vector in a plane is expressed by two numbers. He sought to discover the vectorial force, the fourth proportional, by multiplying the triplet, but to no avail. One day he suddenly grasped a solution to his problem: the geometrical operations of three-dimensional space require for their description not triplets but quadruplets. He formulated these quadruplets and named the set of four numbers "quaternions." "Quaternions" are a strange set, capable of being multiplied as if they were single numbers. Stranger still, they do not behave like ordinary numbers in algebra, for they are "noncommutative." That is, when we multiply three by four we obtain the same product as when multiplying four by three, but in "quaternions," switching the position of the numbers does not yield the same result.

After Hamilton's death his reputation declined, until at the turn of the century, when Whitehead made what was then a bold statement concerning an obscure set of mathematical operations: all the physics known during Hamilton's time could be treated by ordinary algebra, but it could be possible that the discovery of new principles of physics would call for Hamilton's

noncommutative algebra. Some years later, "quaternions," which could be applied to the representation of rotations in four-dimensional space, were used to pattern quantum "reality," especially by Heisenberg, Max Born, and Pascual Jordan. Very significantly, Edmund Whittaker (1954, 52) comments that "Hamilton's formulae of 1843 may even yet prove to be the most natural expression of the new physics."

Now, the remarkable fact is that an imaginary number, $\sqrt{-1}$, was chosen to describe Hamilton's rotations. Imaginary numbers are purely "schizophrenic." They supposedly enjoy no conceivable correspondence to the world, yet they can and have been used to describe the world; hence, in a manner of speaking they have become "real." For instance, it has been discovered that computers can be programmed to solve $\sqrt{-1}$, and, unable to produce an "answer," they begin oscillating between the "1" and the "-1," and will continue to so oscillate, ad infinitum, or until they break down or the power is shut off (Spencer-Brown 1979, xix–xxvii). Here, then, is a purely mental formulation that, in dreamlike fashion, uncannily relates to "reality." A more audacious example of "choice," of condensation, or of "hyperfictionalization" could hardly be found.

The scientist does not attend merely to the buzzing confusion of "brute facts," for if he did, he would get nowhere. What he does, and the only way he can operate, is by a mixture of the "real" and the "irreal," and by exercising a degree of arbitrary choice. For example, if a physicist were asked to justify his use of $\sqrt{-1}$ and Hamiltonian "quaternions," he might point, with some indignation, to the many beautiful equations of quantum theory and how experimental data confirm them—albeit, indirectly. He is not willing to give up his interest in these accomplishments, even though one confronts him with the rejoinder that $\sqrt{-1}$ is "irreal." Physics can be viewed as repeated attempts to place metaphysical disguises upon the face of world process and procedure. After the mask has been worn for a considerable time, it tends to blend with the face, so that one can only with great difficulty see through it. But the mask is ultimately and invariably ripped off, and, the chimera having been exposed, another (partly arbitrary) mask must be called up to replace it. The history of physics is, in essence, a succession of leaps from one fiction to another (Jaki 1978, 198).

Gerald Holton (1973, 289–90) convincingly illustrates Einstein's mixing of "reality" with "irreality." This is also Borges's method, as commented on by some of his critics. Bell-Villada

Interlude (1981, 42), for one, remarks on how Borges's "narrative and its theory deftly combine the physical and the mimetic with the inventive and imaginary; it intermingles the mundane with the miraculous and hallucinatory." Elsewhere (1981, 80), Bell-Villada points out how Borges's exegetical indications faithfully foreshadow the reader's "hesitation effect" theorized by Tzvetan Todorov (1973), in his general theory of fantastic literature. An essential trait of this type of literature is that at certain junctures the reader finds himself unable to discern whether or not the text is "real," factual, or hallucinatory, therefore his hesitation or oscillation. This "hesitation effect," one must acknowledge, must be common also to the contemporary pure scientist, who necessarily constructs a contrary-to-fact conditional, that is, an imaginary construct that entails the choice of a condensed image of model. The scientist then combines this construct with the "real," hoping for an essential correspondence. She must do so because she wants to find out what *would* happen were she not tampering with what is happening. But since she cannot not tamper with "reality," there must exist a certain hesitation, stemming from doubt, that the desired correspondence is itself "real." Hence if she is an honest reader of the "text" of the world, she must admit, perhaps with some embarrassment, that the only "absolute reality" she can speak of is in part subjectively (and arbitrarily) contrived (Wigner 1970, 193). In other words, she cannot help but construct a contrary-to-fact conditional, which is inextricably poetic in nature, as well as, hopefully, cmpirical (Planck 1936, 36).

Contrary-to-fact conditional scientific discourse and fictive discourse share, in addition to explicit or implicit symbols of condensation, the characteristics of a story. This notion has come to the fore in recent years. Jean-François Lyotard (1984) argues that the Western world has been in error over the past couple of millenia, because true "knowledge" is to be derived from narrative rather than from "rational" philosophical and scientific discourse. However, recent arguments have increasingly revealed that scientific theories, beginning as imaginative constructs, *are* narrative, that is, stories, in the sense that they imply a critical and rectifying drama or episode stemming from an effort to determine whether or not these stories are merely stories or—in a fictitious, metaphorical way—somehow about the world. Empirical truth is not the starting point here, as we now know from Einstein and others, but rather, it is merely *a* direction in which scientific inquiry tends to proceed: acts of

mind, thought experiments, and contrary-to-fact conditionals constitute the chief focus of theory-making. In this regard, the scientists' notion of what a good scientific theory should be like is akin to artistic conviction (Medawar 1982, 53; Morris 1983, 64–67). And like storytellers, scientists also avail themselves of unique styles of reasoning, none of which are determined, nor can they be determined, purely by any absolute standard (Hacking, 1985).

More striking yet, Alasdair Macintyre (1980, 54–74) argues convincingly, as does Feyerabend (1975), that "epistemological crises" portrayed in literature and science alike can be viewed through much the same microscope. Macintyre takes pains to demonstrate that authors such as Shakespeare and scientists the likes of Galileo have used figurative language to argue, with shrewd rhetorical moves, for the validity of their viewpoint. For instance, the mythical empiricist account of Galileo tells of his appeal to the facts against Ptolemy and Aristotle. What he actually accomplished was a complete rewriting of the narrative of late medieval science, thus revealing the anomalies that allowed him to generate his alternative interpretations from counterexamples and ad hoc explanatory devices. It then became retrospectively possible "to see how the various elements of various theories had fared in their encounters with other theories and with observations and experiments, and to understand how the form in which they had survived always bears the marks of its passage through time" (Macintyre 1980, 62).

So Ptolemy was not merely pitted against Galileo. New light was shed on the entire tradition, Galileo's predecessors were rewritten and explicated in a different manner, and Galileo's narrative was made to demonstrate why his predecessors must be rejected and modified, and why past theories were erroneous, though according to the narrative of their day they could have remained adequately credible.[4]

It is slowly becoming evident that imaginative scientific constructs, as initiation to theory, *become* narrative, story, fiction. At bottom the two forms of "narrative," science and literature on the one hand, and philosophy and literature on the other, are not entirely unique to one another.[5]

Chapter Four **The Universe as Library**

The physicist Henry Margenau (1949, 250) remarks, in the spirit of Einstein, that "[t]here is something ineffable about the real, something occasionally described as mysterious and awe-inspiring." Those who cease being modern, sophisticated, and ofttimes pedantic seekers of "truth" and become, for a change of pace, old-fashioned wonderers wandering about the world eventually form for themselves a vague notion of the universe, of the totality of consistently interesting things: they are not the victims of a transcendental illusion. This vague notion is necessarily a cosmology, but not in the ordinary sense of the word. It is a general conception of things that can never be determinate and grasped in their totality; it is an open, amorphous, ongoing dialogue with the universe, which is to say that the universe interacts with itself, since the knower, who dialogues with the known, is part of that selfsame known.

However, a desire for the unrealizable—to transcend the given—tends to endure, as Einstein himself testifies: "[I]n a man of my type the major interest disengages itself to a far-reaching degree from the momentary and the merely personal and turns toward the striving for a mental grasp of things" (1949a, 7).[1] In science the paradigmatic form of this "mental

Chapter Four

grasp of things" is by way of the "thought experiment," an act of mind beyond experience as discussed in the Interlude, which, in a manner of speaking, at least simulates one's being outside while remaining within.

In fact, scientific (or aesthetic) creativity conceived in the order of "controlled hallucination" is nowhere more effectively demonstrated than in Einstein's well-known thought experiment that culminated in the space-time continuum, a finite but unbounded system. In this chapter, after briefly introducing Einstein's visionary model—and with an aside to the Eternal Return as Borges formulates it—I map out parallels it evinces with Borges's Library of Babel. If the Argentine writer is not present in certain sections of this and the following two chapters, he remains, nonetheless, ubiquitously lurking in the shadows.

According to one of the Indian schools of philosophy, the ego is merely an onlooker who has identified himself with the man he is continually looking at. The fact that when I write I am stressing certain peculiarities of mine and omitting others has led me to think of Borges as a creator of fancy.
—Jorge Luis Borges

The Fearful Sphere

1 When empirical data are by and large unavailable, such as in microphysics or astrophysics, scientists have often availed themselves of a thought experiment, which, if carried to the limit, can reveal theoretically what the outcome of an initial set of conditions would be. The thought experiment is, in its inception, comparable to constructing an elaborate metaphor, a conceit, or even an allegory—that is, a story—that becomes a condensed symbolic subset of the entire text. The Western rational mind in general has tended to push the imaginative mind, without which reason could not possibly exist, under the rug for centuries. It seems at times that humankind deluded itself into believing the nonexistence of this imaginative mind, but it is alive and well, held together during the imperious dominance of mechanism by a few prophets in the wilderness, including Goethe, who sought to make the imaginary and creative the cornerstone of both sci-

ence and art; Coleridge, who placed it at the center of his philosophical outlook; Nietzsche, for whom myth, the foundation upon which artistic creation must be constructed, takes precedence over logic and reason. Recently, we also have the examples of Bergson and Whitehead from one tradition, Heidegger, Gadamer, and Derrida from another, and, of course, there is the later Wittgenstein.

The Universe as Library

Yet Einstein, a master of thought experiments and perhaps the most creative of twentieth-century scientists, always stressed the idea of a rational and completely cognizable universe. This might appear to be an ambivalent attitude that could stem only from a lingering overconfidence in the power of mathematics. Number, it is well known, ruled the universe of the Pythagoreans. And Plato insisted that the reality and intelligibility of the physical world could be comprehended only through the mathematics of the ideal world. Yet the numbers they spoke of were not numbers in the modern sense of the word. Natural numbers, the integers, reigned supreme. Nor was the universe of the early Greeks a universe transcending immediate sense-perception as is the contemporary scientific view of our universe. It was limited primarily to the concrete. In music, the Pythagoreans saw verification of number theory, and sight and touch were expressed in their geometrical configurations. They relegated the unthinkables, the unutterables (e.g., irrational numbers, infinity, etc.) to forgetfulness (Dantzig 1930, 99–111).

Modern mathematics transcends the Greeks' limited sense of number. The unthinkables have been extracted from the closet and used to describe a hypothetical world inaccessible to the senses. Surprisingly, Einstein's imaginative thought experiments, by his own admission, are in general about entirely visual events, e.g., catching up with a photon of light or standing in an elevator either being accelerated upward or in a state of free-fall. In contrast, other physicists' thought has followed nonvisual cues. Arthur I. Miller (1978) gives a fascinating account of the history of quantum theory from 1913 to 1927. There was a struggle between two camps. On the one hand, scientists such as Bohr, Planck, and Schrödinger preferred visualizable models (wave mechanics); on the other, Heisenberg and Max Born found wave models "disgusting" and wanted them replaced by, among other abstract mathematical models, matrices and groups. These two antagonistic coalitions, after a few years of bickering, arrived at an uneasy general consensus, the so-called Copenhagen interpretation, according to which

Chapter Four   the wave model and the abstract (particle) model are complementary, each enjoying its share of "reality." Miller concludes that the general scientific mind requires a variety of modes of thought, as does the artistic mind.

Borges, interestingly enough, was never a visual person, unlike, for example, the Chilean poet Pablo Neruda, whose verbal pyrotechnics create concrete imagery in the reader. In Borges's best stories, the visual is subordinate to mental patterns. However, the visual element is never entirely eradicated. The Library of Babel or the Aleph are strictly mental in their allusions to infinity, Cantor's sets, and Zeno's and other paradoxes. And they are also visual. A section of the Library provides a concrete image, though its potentially infinite extension does not. The Aleph as a small, visible sphere is concrete; its relationship to the entire universe is the most severe of abstractions. In "Death and the Compass," references to colors, sections of the city, and the shapes of windows, rooms, stairways, galleries, and gardens are visual; the symmetries of space and time, and the vague allusion to Zeno, are not. Suffice it to say that Borges realizes an almost uncanny convergence of the abstract and the concrete. Indeed, he remarks at the conclusion of "Avatars of the Tortoise," one of his most abstract pieces, though it is not devoid of imagery, that "[i]t is hazardous to think that a coordination of words (philosophies are nothing else) can have much resemblance to the universe. It is also hazardous to think that one of those famous coordinations does not resemble it a little more than others, even in an infinitesimal way" (*OI*, 114).

Borges goes on to assert that the only person in which he could recognize some vestige of the universe was Schopenhauer, who asserted that the world is a fabrication of the will. To repeat, Borges asks us to admit what all idealists admit, that "the nature of the world is hallucinatory," a tall order indeed. But Borges also gives us an even more arduous task, to "do what no idealist has done: let us look for the unrealities that confirm that nature." The "unreality" of art, Borges suggests, can become an instrument for discovering the idealist nature of the world. The scientific thought experiment, as constructed by Einstein and others, is an "unreality" that can, by interpolation into the world of experience, become provisionally "real," a "reality" that is ultimately tantamount to the "unrealities" of art. In this context, there is no discernible line of demarcation between scientific and artistic worlds. A continuum bridges what is ordinarily presumed to be an abyss.

The Universe as Library

Emir Rogríguez Monegal (1978, 121) tells of a piece by Borges in the magazine *El Hogar* (11 June 1937, p. 30), in which he refers to a discussion with the Indian poet Rabindranath Tagore about Baudelaire's "Death of the Lovers," a sonnet cluttered with concrete images (divans, beds, flowerplaces, mirrors, shelves, and angels). Tagore listened intently, and finally said: "I don't like your furniture poet!" Borges reveals, understandably, that he sympathizes with Tagore, for he too is a poet with a propensity for "unreality," for an unlikely fusion of the "real" and the "irreal." Einstein also once had a conversation with Tagore in which he defended his notion of "reality" by way of a thought experiment, as if there were no absolute observer and yet as if Einstein himself were beyond the universe. Tagore patiently responded that even though such a "reality" might exist, it would in all probability not be accessible to the scientist; he would not know how to speak in intelligible terms about it (Buckley and Peat 1979, 73). Tagore was obviously referring to a "reality" available only through mystical experience. Yet Einstein persisted in his efforts to acquire a mental grasp of that "reality" by means of a contradictory mixture of the "real" and the "irreal," of reason and unreason, as does Borges. This is, precisely and perennially, one of the chief quandaries of the Western mind: to comprehend the unintelligible, to describe the indescribable, by sheer intellection.

It is common knowledge that in contemporary times some of the most bizarre and even outlandish narrated events are to be found in science fiction, which, common knowledge tells us, entails a "real-irreal" amalgamation par excellence. Science fiction, however, must take a backseat to contemporary physics. John Gribbin illustrates in *Time-Warps* (1979), while citing science fiction writers as well as scientists, that the "new physics" is even stranger than the strangest science fiction. In fact, science fiction writers are now borrowing ideas from physicists, rather than, as has been conceived in the past, science "catching up" with H. G. Wells, Jules Verne, and others. Perhaps this is in part because many of the "new physicists" are mathematicians first and observers of empirical "reality" second. The pure mathematician does not concern herself with the world, but with thought. Her constructs "are not only created by thought but consist of thought, just as the creations of the engineer consist of engines" (Jeans 1930, 145).

One of the products of pure thought, a combination of the "real" and the "irreal" conceptualized in relativity theory, is the idea that the universe is finite but unbounded. This idea is not

Chapter Four

exactly novel. In "Pascal's Sphere" (*OI*, 6–9), Borges briefly recapitulates the long history—which, he says, is the history of a few metaphors—of the notion of finite but absolute space, beginning with Lucretius and Xenophanes and terminating with Pascal's "frightful sphere, the center of which is everywhere, and the circumference nowhere" (*OI*, 9). He exemplifies this venerable concept with his Library of Babel, which is called the universe and which has "an indefinite and perhaps infinite number of hexagonal galleries" (*L*, 51). In fact, the narrator of The Library paraphrases Pascal with the classic dictum: "*The Library is a sphere whose exact center is any of its hexagons and whose circumference is inaccessible*" (*L*, 52).

How is it possible to verbalize this "ineffable" concept? Early relativity physicists lecturing on infinite and unbounded space—the product, after all, of a thought experiment—were usually met with the objection that the concept is self-contradictory and absurd. If space is finite, it must be possible to go outside it, therefore there must be something outside it, but there cannot be, so the idea is nonsensical. To this argument Jeans (1930, 156) responds that these critics remained imprisoned within their nineteenth-century scientific state of mind, for they assumed the universe to be intelligible by "material representation." Jeans continues:

> If we grant their premises, we must, I think, also grant their conclusion—that we are talking nonsense—for their logic is irrefutable. But modern science cannot possibly grant their conclusion; it insists on the finiteness of space at all costs. This, of course, means that we must deny the premises which our critics unknowingly assume. The universe cannot admit of material representation, and the reason, I think, is that it has become a mere mental concept.

After various allusions to Berkeley, Jeans then waxes rather mystical, arguing that if the universe is a universe of thought incompatible with mechanistic descriptions, then its creation must likewise have been a mental performance—i.e., Jeans's "Great Mathematician." In fact, the very finiteness of time and space oblige us to think of creation as such an act.

This reference to the creation is apropos. It is believed by most astrophysicists that before the "big bang," which initiated the evolution of the universe, all matter was contracted to an infinitesimal point. Upon hearing this inconceivable statement,

one naturally jumps to the conclusion that the point was surrounded by space, but it was not, nor was it surrounded by a void. This is the problem of conceiving the universe somehow to be a sphere. A sphere has a surface and a center, but the universe, we are told, has neither. The problem with this formulation is that we risk becoming bogged down with infinity and Zeno's paradoxes once more.

In fact, the apparent absurdity of the notion of finite but unbounded space derives from Lucretius, who resorted to a thought experiment something like this: Suppose space to be a gigantic sphere, and someone travels close to its edge and throws a dart at it. Either the dart can penetrate it or not. If it does, then space has no real boundary; if it does not, then there is nothing outside and therefore still nothing to stop the dart (Rucker 1983, 16–18). The revulsion of the Greeks toward such a notion of unboundedness was such that Parmenides, Plato, and Aristotle all believed space to be bounded and finite. Finitude offers the security of closed space, but there persists the feeble hope of somehow being able to break out; infinitude, however, is a prison from which there is no escape. Borges's description of the Library is even more fearless: it is "unlimited and cyclical. If an eternal traveler were to cross it in any direction, after centuries he would see that the same volumes were repeated in the same disorder (which, thus repeated, would be an order: the Order)" (*L*, 58).

Finally, the finite but unbounded can be metaphorically illustrated with one of Escher's prints (see figure 3).[2] There are an infinite number of ways of interlocking the "heavenly beings" and the "hellish beings," which diminish in size as they radiate out toward the circumference. This type of design has been called a "Poincaré plane" after the mathematician Henri Poincaré, who suggested that it might offer a pattern of a finite universe with an unattainable boundary (Barrow 1990, 317–21). The circle, which is finite, can, like the Library, represent a sphere. It is an unbounded sphere in the sense that as the figures approximate the perimeter they become more numerous, yet no matter how large their number becomes, they will never exactly merge into the boundary line—which, of course, must be nonexistent, following Zeno. Such an image approaches de Cusa's polygon with an infinite number of angles, the circumference of which, Borges (*OI*, 109) tells us, would become a line, a triangle, a circle, a sphere, etc., which is also the image of the Library. If each of the Library's galleries is hexagonal, and if there is an indefinite

Chapter Four   number of hexagonal rooms, perhaps an infinity of them, then the universe (i.e., the Library) will consist of a sphere of an infinity of hexagonal faces. We must be mindful, however, that the sphere is unlimited and cyclical, a disorder which, when repeated, hopefully becomes the Order. One might contend, however, that Escher's print and Borges's Library must have a center: Escher's, since the figures grow larger as they approach the middle point, and Borges's, for if all the galleries are of equal size, as he describes them, then an eternal traveler has only to count the hexagons to determine the sphere's diameter, radius, and central gallery. The problem is that he who gazes at the print cannot locate its exact periphery, so the center is unavailable to him. And the Library's eternal traveler is presumably infinite, hence any center must be inconceivable for the finite being.

Circle Limit I. 1958. © M.C. Escher/Cordon Art—Baarn—Holland

Figure 3

Paul Davies (1981, 130–50) provides a tropological and topological image for the finite but unbounded universe. Consider the universe to be like a vast cylinder, the curvature of which represents our three-dimensional space and the longitude of which represents one-dimensional time (see figure 4; also Davies 1981, 135). A snapshot of this universe would be like taking a slice out of the cylinder at a given point. This slice would be a circle representing our finite but bounded three-dimensional space. Along such a circle a traveler could begin, pass through the entire universe, and return to his starting point from the opposite direction; or, with a powerful enough telescope a superastronomer could supposedly gaze along the curvature of space and see his backside. Notice that this slice is finite, yet there is no boundary, no center, no edge—keeping in mind that the periphery itself represents space, for the slice does not enclose a space; that is, the region inside the cylinder is not part of the universe, it is used only for the convenience of a visual model.

The Universe as Library

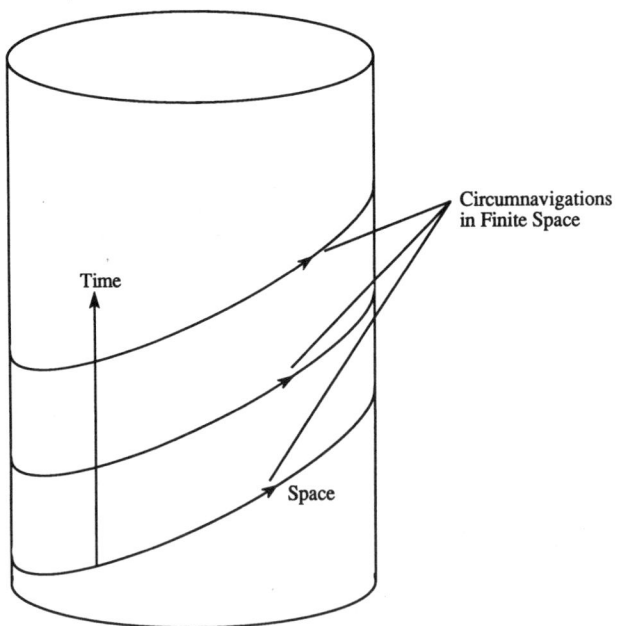

Figure 4

Of course, for us to conceive of a static slice from this cylinder-universe is equally as impossible as the task Laplace gives his superobserver, since time intimately embraces space in

Chapter Four   a four-dimensional manifold. In this case, the traveler would not actually trace out a circle but a helix, for bound within time, his voyage would not bring him back exactly to the beginning, but at a future time to a point further along the cylinder. Our picture still remains inadequate, however, for the universe is expanding. In such case, beginning at the infinitesimal point where the "big bang" occurred, space begins its expansion through time, which can be represented by a cone (see figure 5; also Davies 1981, 138). As time proceeds along the cone from apex to base, the galaxies move apart in space, so that to a given observer at any point on the cone it appears that the galaxies are receding from him as if he were at the "center" of the universe: *his* universe. Unfortunately this is also a somewhat false appearance, for as expanding space drives the galaxies apart, they do not move away from any given observer; yet insofar as each observer is concerned, they are moving away from *him*. Any observer on any planet, then, can consider himself to be at the "center" of the universe. (Einsteinian relativity is perhaps the most "democratic," and yet the most "egocentric," cosmology ever invented.) One might initially conclude that Borges's Library, like Pascal's sphere or Escher's print, is static rather than expanding, and since it is susceptible to Zeno's dilemma, it is also an inadequate hypostat of the finite but unbounded universe. As we shall note below, however, in its composite form the Library is more faithful to the general twentieth-century cosmology than immediately meets the eye.

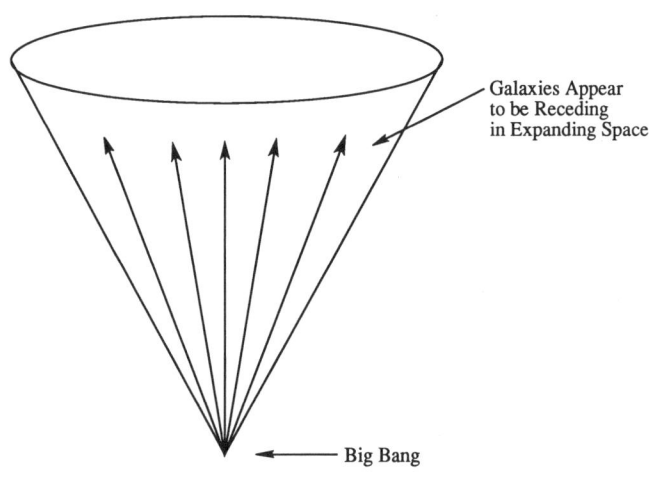

Figure 5

The Universe as Library

With the notion of a "cylindrical," or better, a "conical" and expanding universe, I have indirectly alluded to the Eternal Return, which never ceased to fascinate Borges. Let us briefly consider this concept, since it will have an indirect bearing on future discussions. Then I will return to the Einsteinian space-time continuum. Borges alludes to cyclical time often. His poem "The Cyclical Night" (*OP*, 182) begins and ends with the same line: "It was known by the arduous pupils of Pythagoras." In certain short stories, namely "The Library of Babel," "Tlön, Uqbar, Orbis Tertius," "The Immortal," and "The Garden of Forking Paths" (discussed in the third section of Chapter Six), the theme appears often. Borges also dedicates two essays, "The Doctrine of Cycles" (*HE,* 75–89) and "Circular Time" (*HE,* 91–97) to the Eternal Return, with reference to Nietzsche. Borges confessed to his once having worked on a mathematical notion of endless time; in such a situation everything would eventually happen; this concept later became the theme of "The Immortal." But the question is, would, or indeed could, everything happen?

By his own admission, Borges as a writer is caught up in an infinitely repetitive regress, for he remarks: "I tend to return eternally to the eternal return" (*HE,* 91). And he once told an interviewer: "I'm always writing the same story; I have three or four arguments for stories, but then I give different treatment to these three or four stories, I write them with different inflexion, I situate them in different periods, in distinct circumstances: and then, they become new" (Peicovich 1980, 111). In a passage quoted above from "The Library," Borges makes reference to the eternal traveler who eventually sees the same volumes repeated in the same order. This is one of the most obvious examples in his prose of Nietzsche's interpretation of the eternal return. Nietzsche uses this concept to critique science, specifically the mechanistic corpuscular-kinetic science of the nineteenth century:

> If the universe may be conceived as a definite quantity of energy, as a definite number of centers of energy—and every other concept remains indefinite and therefore useless—it follows therefore that the universe must go through a calculable number of combinations in the great game of chance which constitutes its existence. In infinity, at some moment or other, every possible combination must once have been realized; not only this, but it must have been realized an infinite number of times. And inasmuch as between every one of these combinations and its next recurrence every other possible combination would necessarily

have been undergone, and since every one of these combinations would determine the whole series in the same order, a circular movement of absolutely identical series is thus demonstrated: the universe is thus shown to be a circular movement which has already repeated itself an infinite number of times, and which plays its game for all eternity. (Nietzsche 1913, 430)

For Nietzsche a universe of chaos, of random chance, often believed to be the essence of the Newtonian worldview, excludes purpose. However, it does not contradict the notion of cycles, for this notion is only an irrational necessity. Chaos and cycle, becoming and the eternal return, are usually conceived of as oppositions, but this is not correct, says Nietzsche. Borges has often toyed with the idea that even though literature (or language) is inexhaustible in terms of all possible combinations, the eternal return remains to haunt one. He begins "The Doctrine of Cycles" by demonstrating some knowledge of science—which he has denied in interviews—then he turns to a discussion of the notion of a world consisting of ten atoms and their combinations:

> How many different states can this world know before an eternal return? The investigation is easy: It is sufficient to multiply $1 \times 2 \times 3 \times 4 \times 5 \times 6 \times 7 \times 8 \times 9 \times 10$, a tedious operation that gives us the sum of 3,628,800. If an almost infinitesimal part of the universe is capable of this variety, we should allow little or no faith in the monotony of the cosmos. I have considered 10 atoms; to do the same with 2 grams of hydrogen, we would need much more than a billion billion atoms. (*HE,* 76)

Borges remains unconvinced at this point. He then discusses Cantor's set theory argument (*HE,* 77–78, also *D,* 118–20) remarking that with Cantor, Nietzsche is defeated, since if there is an infinity of points in the universe, then the number of combinations is infinite. On the other hand, many of Borges's stories, especially "The Aleph," imply the Cantorian thesis that each subset contains an infinity of combinations, for if "reality" possesses the characteristic of a dense continuum, then, like a line, each segment contains as many points as the whole. Borges, nevertheless, gives Nietzsche his due respect, for he "followed a heroic path. "Nietzsche dug up that intolerable Greek hypothesis of Eternal Return and attempted to infer from it an occasion for jubilance. From the most horrible idea possible, he proposed that it be the focus of delight (*HE,* 82–85). Nietzsche wanted humankind to be capable of carrying eternity on its

weary shoulders, and in consolation, he promised us our immortality through the eternal return.

Borges also alludes to Nietzsche's reference to the second law of thermodynamics, claiming that it actually cancels the "circular labyrinth" of the eternal return with the entropy principle. Entropy, the evolution of the universe from a state of lesser probability to one of greater probability, will eventually reach an end: heat death. Then, after summarizing Plato's and Nietzsche's interpretations of the eternal return, in "Circular Time" Borges outlines yet another dilemma, which is "the least frightening": cycles are similar, but not identical (*HE*, 94).

The "cone universe" discussed above demonstrates essentially the same. Without necessarily recurring to this image, some further computations can bear out Borges's thesis that the eternal return is, for practical purposes, "unthinkable," though it is describable in mathematical language. Let us imagine a variant of "Maxwell's demon," that most worthy candidate for an encyclopedia of imaginary beings capable of reversing the direction of entropy. Suppose we have a box of gas molecules with a partition in the middle and a trapdoor in the partition barely large enough to allow a molecule of gas to pass through. The demon possesses superreflexes enabling him to gather information concerning the velocity of each molecule as it approaches the trapdoor. He proceeds to open the door for the fast ("hot") molecules but not for the slow ("cold") ones, until his side of the box has warmed him comfortably, while the other side has become unbearably cold. He has, in other words, produced the least probable state of affairs; he has brought about a relatively ordered state from a relatively disordered one. This reverse process has often been called "negentropy."

Using set theory, David Park (1980, 52–56) illustrates what the results of this industrious demon might be. Assume a world of two particles. At any given moment there would be four possibilities: both molecules on the left side of the box, both on the right side, molecule a on the left and b on the right, or molecule b on the left and a on the right (if the two molecules were removed from the box it would be the equivalent of the "null set"). For three molecules the state of affairs would entail $2 \times 2 \times 2 = 8$ possibilities, for four, $2 \times 2 \times 2 \times 2 = 16$ possibilities, and for N molecules there would be 2^N possibilities. According to this formulation, before we might find N molecules on a particular side of the box we would be required to

look a maximum of 2^N times at our experimental contraption. Now let us assume we space our observations 1/100 second apart—an impossibility, of course, for the plodding human organism. With a two-molecule system we would be required a maximum of 4/100 seconds to find the molecules on a given side, and with 2^N molecules the total time required before finding all of them on one side of the box would be T = 1/100 × 2^N seconds. Park then provides the following computations:

Number of molecules	Time (= 1/100 × 2^N)
10	10 seconds
20	10,500 seconds = 3 hours
50	350,000 years
100	4×10^{20} years

In order to grasp the magnitude of this last number, the age of the universe has been estimated at 2×10^{10} years. If we divide this number into the number of years for a particular combination of 100 molecules to repeat itself, we find that it is 2×10^{10} times as large! Since the total number of electrons in the universe has been estimated at a mere 10^{79}, I will leave the calculation of an instance of the eternal return to the enterprising reader. Theoretically the eternal return is valid, but this fact should give us no cause for alarm, not yet at least. In this light, Borges's skepticism of Nietzsche is understandable.

Throughout his work, Borges reveals two themes that he reiterates at the end of "Cyclical Time": (1) universal history as the history of one man (or conversely, the combined lives of all men is one man)—that is to say, there is "nothing new under the sun"—and (2) the Schopenhauerian affirmation of an ever-flowing present. The first is Parmenides, the second Heraclitus. Both, of course, have been predecessors to an illustrious line of thinkers; one or the other of their ideas has periodically enjoyed prominence throughout history. Borges then concludes on a rather positive note: "In heightened times the conjecture that man's existence is an invariant, constant quantity, can depress and irritate; in declining times (like these), it is the promise that no ignomy, no calamity, no dictator can impoverish us" (*HE*, 97).

Following his customary strategy, the Argentine man of letters finds a way to use any and all ideas, even the most

undesirable ones, to his aesthetic and personal advantage: a disbeliever who, nonetheless, cannot free himself from his disbeliefs, who is incessantly encumbered by them. Let us now direct our attention toward a distinctly Parmenidean conception of the universe, which at least Einstein and a few like-minded thinkers have believed to be the one and only "reality."

The Universe as Library

> *I have come to believe, though very reluctantly, that [mathematics] consists of tautologies. I fear that, to a mind of sufficient intellectual power, the whole of mathematics would appear trivial, as trivial as the statement that a four-footed animal is an animal. I think that the timelessness of mathematics has none of the sublimity that it once seemed to me to have, but consists merely in the fact that the pure mathematician is not talking about time.*
> —Bertrand Russell

2 Pedro Amaral (1971, 426–27) remarks that Borges's Library evinces parallels with Parmenides and the Einsteinian finite but unlimited universe. This view is to be applauded, for an effective reading of Borges's metaphysical prose demands an approach both general and penetrating, encompassing a broad vision, but without the shallowness endemic in many interdisciplinary approaches. On the other hand, Ion Agheana (1984) in general emphasizes that Borges's interests regarding space and time are chiefly literary; hence his metaphysical asides should be viewed as creative embellishments. I cannot disagree, from a strictly literary point of view. For Borges the writer, not the thinker who takes occasional liberties to scrutinize various metaphysical perplexities, time and space are the product of accumulated human experience. However, while harboring no disregard for exclusively literary studies of Borges, the task I have set for myself is obviously something other: to elucidate the concepts embodied in the texts of a writer whose acuity of insight is exceeded by few, if any, contemporary artists. It is my hope that the present chapter does justice to the depth and wisdom of this insight.

Parmenides Triumphs

Chapter Four

In the preface of *The Book of Imaginary Beings,* Borges begins: "The title of this book would justify the inclusion of Prince Hamlet, of the point, of the line, of the surface, of *n*-dimensional hyperplanes and hypervolumes, of all generic terms, and perhaps of each one of us and of the Godhead. In brief, the sum of all things—the universe" (*BI,* 13). Here we have another strange juxtaposition that tends to engender an uneasy feeling. Why are mathematical constructs thrown into the same heap with a character from a play, ourselves, the Godhead, in short, everything? What generic classification is capable of encompassing this totality? It seems that the only answer, to echo Nietzsche, is: Fictions All!

Judging from Borges's passage and from many of his other bizarre connections, juxtapositions, and classifications, everything in the universe is coequal; nothing enjoys priority or privilege over anything else, all things are contiguous with all other things. Humankind has suffered such disappointment, frustration, and anguish because it persists in nurturing a hope for answers that cannot be forthcoming; it is incessantly driven by purposes, goals, visions of the sublime; its mental set remains chiefly teleological in orientation—Faustian man. Teleological expectations inevitably run against a brick wall. Humankind exists within, not outside, nature, and nature *is* as it *is;* it simply exists, without purpose. A nonteleological framework is essentially "democratic." It abolishes the Romantic idea of individual genius. There is no creativity *sui generis,* no discovery, but merely invention, which entails a recombination of what there is. Significantly, Borges has repeatedly denied that he has anything new to say, and this come from the mouth of an extraordinary writer. If "reality" cannot be adequately described without using fictive points and lines, $\sqrt{-1}$, references to hypercubes (in relativity theory), etc.; if the godhead is yet another fabrication of our desires; if imaginary beings can be non-ontological "objects," then of what can the furniture of the world possibly consist? Can it indeed be "real"? If we extrapolate the premises of Borges's stories to the extreme, we are forced to conclude that mental worlds are invariably interjected into the nonmental sphere, and traditional dividing lines between "real" and "irreal" become riddled with gaps and leaks. All-or-nothing distinctions can be no more than wishful thinking.

Borges's interpenetration of the "irreal" into the "real" and vice versa is prominent in various stories. The imaginary Tlön overtakes our world. In "The Lottery of Babylon" (*L,* 30–

The Universe as Library

35) (to be discussed in the second section of Chapter Six), a game of chance takes on increased power until it becomes the only reality. The entire universe is sucked up into a sphere about the size of a golf ball—the Aleph. The world of "The Immortal" becomes lost in eternity. In "The God's Script," one cipher is capable of replacing the universe. The narrator's world of "The Zahir" gives way to an insignificant coin. And, to climax these supplantations, many believe the Library to be The Universe: the world as a text. Borges longs for the view *sub specie aeternitatis,* as did the early Wittgenstein, also an admirer of Schopenhauer. But, according to the premises put forth in this study and following Borges's own occasional disclaimers, such a panoramic grasp is out of the question. The fact remains, however, that Borges, like Einstein and others, continued his efforts to realize the impossible by postulating purely imaginary constructs "as if" they were themselves "outside."[3]

In the present century, the mind act par excellence is Einstein's convolution of space and time into a monolithic four-dimensional manifold. Einstein, however, was not exactly a prophet in the wilderness. The idea of a fourth dimension was definitely "in the air," and, it bears mentioning, the idea never ceased to captivate Borges. For example, during the revival of mysticism around the turn of the century, some thought the fourth dimension to be the ideal candidate for the source of all mystical experience. Chief among the books written in that period was Hinton's *The New Era of Thought* (1888). Hinton claims that it is possible to acquire a mental grasp of a dimension of space higher than that in which we believe we live. He provided in his book a series of mental exercises along with twenty-seven three-dimensional colored cubes that fit together into a single cube, which would supposedly be four-dimensional, a "hypercube." If one could learn the relations between all the cubes, it was claimed, then one could know the fourth dimension.[4]

Borges, as mentioned above, knew Hinton's work,[5] and, though he is notorious for laying bare the absurdities and inconsistencies in any and every conceptual system, he never ceased harboring a desire to encompass the totality mentally, nor did he properly take Hinton to task for his grandiose claim. It is not surprising, then, that many of Borges's stories suggest a dimension higher than that of our commonsense world. Labyrinths, for example, fascinate him. Perhaps one reason is because from a higher perspective, the labyrinth presents no problem. A traditional two-dimensional labyrinth or maze can

be resolved from a higher view, in the third dimension. At another level, the mansion at Triste-le-Roy in "Death and the Compass" represents a three-dimensional labyrinth, which presumably could be deciphered from a four-dimensional perspective. As will be noted in the third section of Chapter Six, the temporal labyrinth in "The Garden of Forking Paths" can be resolved from within a four-dimensional space-time manifold. There is no absolute labyrinth short of infinity, and conceivably, though paradoxically, from an infinite perspective any and all finite labyrinths can be resolved.

The Zahir and the Aleph can be used to symbolize this relationship between dimensions. The former is in a sense two-dimensional, affording no more than a succession of perspectival glances. In contrast, the Aleph is three-dimensional, incorporating all possible perspectives in simultaneity. But this is, logically speaking, impossible, for to perceive everything in three-dimensional space, the fourth dimension is necessary. However, the narrator, a three-dimensional being, is prohibited from entering the fourth-dimension, so how could he perceive it? One might respond that the Aleph is merely a fiction and that we should pay this anomaly no mind. On the other hand, the Aleph can provide insight into an important Borgesian strategy. It is often remarked that the imagination is incapable of apprehending the fourth dimension, though it can be described mathematically. Some mathematicians, with the aid of computer mock-ups, report that though they can neither visualize four dimensions nor describe them in natural language, they can get a "feel" for the hypercube and other such "objects."[6] Mathematics entails strict formal reasoning, while reasoning based on visual models and figures like the hypercube is intuitive. By intuition, a model or figure can be conceived and then given formal dress. The Aleph is such an intuitive model. In fact, its counterpart, a "space-time singularity" (to be discussed in the second section of Chapter Five), is precisely a scientific model that has been described mathematically within the context of the four-dimensional space-time manifold.

To establish the relationship between the Zahir and the Aleph more adequately, suppose the Zahir to be an inhabitant of Flatland and the Aleph an inhabitant of a three-dimensional realm.[7] Zahir's perceptual faculties are, of course, limited to two dimensions, back and forth along a plane. Were he to see a triangle with its apex pointing toward him, it would appear as a line that would be distinct in the middle and become blurred at its

extremities. Aleph is more fortunate, for from above, within his three-dimensional world, he can perceive that Zahir is round, and that there are squares, triangles, ovals, and a host of other shapes in his world. Imagine Zahir's surprise when Aleph decides to pass through his flat world. Aleph will appear in Zahir's perceptual field first as a point, then as a short segment that is slightly blurred at the extremities. As the segment grows longer, and after Aleph's circumference has passed through Zahir's planar world, the process reverses itself until a mere point shows, which quickly vanishes. (If you imagine the surface of a swimming pool to be Flatland and pass a beach ball through it, the effect is the same.) Zahir ordinarily possesses no capacity for perceiving Aleph as a three-dimensional being, for he is physically trapped within his own world. By means of a thought experiment, however, he can potentially create the concept of up-and-down motion, but to do so he must imagine himself outside his own realm of existence; this, for him, is an exceedingly difficult task. The problem is that even if he were able to jump out of his Flatland, his vision would remain two-dimensional, hence the most he could do is mentally scan his world a slice at a time, like the "dot" that scans the TV tube or like the slice out of the imaginary cylinder-universe discussed above. In contrast, Aleph, enjoying three-dimensional vision, can see an entire patch of Zahir's world at once.[8]

This brief digression emphasizes further that a higher dimension can neither be visualized nor described in natural language, but it can be conceived and described mathematically—as Einstein and relativity physicists have demonstrated. Borges's stories are in this sense, as has been implied, chiefly concept-laden rather than image-laden. Even though the narrator of "The Aleph" might be able to transcend his three-dimensional existence, his perception would remain like that of Zahir, a mere "slice" out of the whole at each moment. Most significantly, the narrator of "The Aleph" confesses that words are inadequate tools for describing "the limitless Aleph, which my floundering mind can scarcely encompass" (A, 26). This must be the case, for, regarded as a point in four-dimensional space-time, the Aleph can be neither visualized nor described.

In relativity physics, the impossibility of separating space and time was first proposed by the Russian mathematician Hermann Minkowski in 1908, three years after Einstein's special relativity appeared. Shortly thereafter, the concept of a relativistic four-dimensional space-time continuum, "Minkowski space,"

a space in the true mathematical sense of the term, was conceived. This is not to say that the space-time continuum is as if space were four-dimensional, though it has been so misrepresented by the hypercube, which is no more than a rough visual model constructed in an attempt to illustrate the unvisualizable. More satisfactorily stated, three-dimensional space and one-dimensional time are intimately interwoven in relativity theory, and they are simply meaningless if separated; distinct models of each cannot be adequately constructed. Before the Einsteinian revolution, distances and time were presumed to have an invariant value for all observers. Since it was tacitly assumed that light travels at infinite velocity, the notion of simultaneity persisted. If one thousand observers on one thousand different planets were to take a photograph of our sun simultaneously, those photographs would supposedly reveal slightly different perspectives of the universe "at the same instant." Laplace's Superobserver was thus made possible, that omniscient being who, aware of the position and velocity of every entity in the universe, could predict all future states and compute all past states. In Einsteinian space-time, in contrast to this Newtonian framework, each observer carves out his/her own space-time. And since the velocity of light is finite, each observer therefore perceives a slightly to greatly distinct universe. There can be no simultaneities.

Minkowski space has been called the "block" universe. In the words of Louis de Broglie (1959, 114):

> In space-time, everything which for each of us constitutes the past, the present, and the future is given in block, and the entire collection of events, successive for us, which form the existence of a material particle is represented by a line, the world-line of the particle. Moreover, this new conception defers to the principle of causality and in no way prejudices the determination of phenomena. Each observer, as his time passes, discovers, so to speak, new slices of space-time which appear to him as successive aspects of the material world, though in reality the ensemble of events constituting space-time exist prior to his knowledge of them.

Each particle (or at a macroscopic level, each observer) is a "wanderer" along his unique world-line. A given observer within the Minkowski model cannot enjoy the same panorama as his distant cousin, for she is trapped within her own light cone (see figure 6). A particular light cone exists for the path of only

one particle-observer (P-O). The circumference of the base of the two cones depends upon the constant velocity of light and hence can be broadened or narrowed depending upon P-O's velocity. The outside ("elsewhere") is not accessible to P-O, though a portion of the "elsewhere" might be accessible to his neighbor within her light cone, but at the sacrifice of her own "elsewhere," a portion of which in turn might lie within P-O's light cone. Light cones necessarily divide space into two regions, outside and inside. P-O's world-line must always remain on the inside because it moves at a velocity slower than light. If this model were projected into Newtonian space, time, and the infinite velocity of light, the two cones would flatten out to a horizontal plane; P-O would now potentially be omniscient—Laplace's Superobserver.

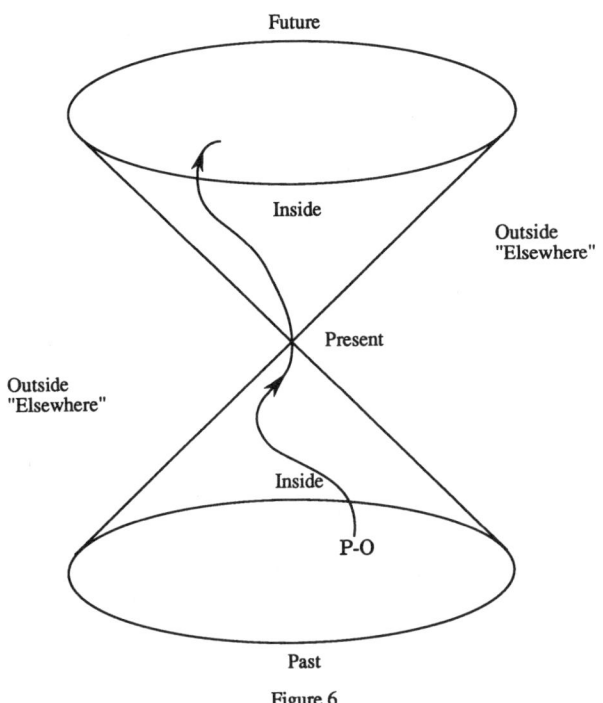

Figure 6

Assuming P-O's world-line as well as those of all his relatives to exist in an inconceivably massive cube, we now have a rough imaginary model of Minkowski space, the "block." This totality simply *is,* it does not become. It appears to become only to the "traveler" along his world-line (Weyl 1963, 194). Since the

Chapter Four "block" in its entirety is not available to P-O, this "traveler" has no recourse but to resort ultimately to imagination-conception rather than perception in his effort to comprehend the universe. And such imagination-conception is made possible by means of conjectural leaps, in the manner of Einstein, who was able to imagine himself "as if" on the outside, or Zahir imaginarily jumping out of his two-dimensional world. In spite of the strange features of this "block" universe, few physicists now seriously doubt the theory of relativity.[9] The "block" is like a sheet of music directing that all the notes be played simultaneously with one great crash, a novel which must be read in an infinitesimal fraction of second, or the Buddha, who in a flash sees the arrow leave the bow, arc through the air, and strike the target.

Borges's Library and the "block" universe enjoy common features. There are basically three modes of conceiving the Library: (1) the hexagonal structure, (2) the books housed therein, and (3) the Library's inhabitants. These correspond to three familiar domains: (1) the inorganic, (2) the informational, and (3) the organic. Critics have often commented on the Library's beehive-like anatomy. It is also the perfect example of crystalline form—i.e., the inorganic domain. The ions making up a crystal lattice according to the intensity of their electrical charge and their radius accommodate themselves spatially into the most economical form possible. Such inanimate crystallization processes, like the Library, terminate in static patterns. In fact, if the "mother liquor" in which a crystal is forming is approximately 100 percent pure, the crystal will approach perfect symmetry. Few objects of nature evoke more romantic sentiment and fascination than the masterfully cut gem. Their beautifully designed patterns are a wonder to behold. But they are not properly of our world. Belonging to the inorganic realm, they are, because of their very perfection, static, representing order, death—eternity.

The pristine beauty and harmony provided by the Library's hexagonal structure and the "block" universe, one of the culminating visions of Western science, are analogous "hyperfictions." Both are absolute perfection carved out of chaos: both *are* as they *are,* static, timeless. All of Borges's travelers in the Library describe, during their lifetime, nothing more nor less than an almost infinitesimal world-line within the vast geometrical edifice. As Borges remarks elsewhere (*OI,* 128): "The steps a man takes, from the day of his birth to the day of his death, trace an inconceivable figure in time." But the Divine Intellect,

from its eternal perspective outside the universe (the Library or the "block"), "perceives that figure at once, as man's intelligence perceives a triangle." Or, I might add, as Aleph perceives Zahir, or as we perceive a maze from above.

The enigmatic mirror in each hallway of the Library provides another connecting link to the notion of a "block" universe of space-time continuity. The mirrors, faithfully duplicating all appearances, project beyond the Library's anatomical structure, such projection necessarily entailing a higher dimension. For instance, the mirror reflection of a symmetrical hexagon produces an identical image. A three-dimensional asymmetrical object placed before a mirror, on the other hand, creates an inverse image called an enantiomorph. Enantiomorphs are common in molecular structure. Sucrose, tartaric acid, and a host of other molecules can exist in left-hand or right-hand mirror-image varieties. They are comparable to right-hand and left-hand gloves, each of which, when placed before a mirror, becomes the other. Enantiomorphic forms were puzzling to Kant. He writes in section 13 of *Prolegomena to Any Future Metaphysics* (1950): "What can more resemble my hand or my ear, and be in all points more like, than its image in the looking glass? And yet I cannot put such a hand as I see in the glass in the place of its original." This and other enigmas led Kant to his *a priorist* conception of space.

Enantiomorphic forms also bear on the above discussion of dimensionality as well as on the nature of the four-dimensional space-time manifold and its relation to our three-dimensional empirical world. In brief, an intriguing method for creating an enantiomorph is with a Möbius strip. Assume the flatlander, Zahir, to inhabit a "Möbius strip world." He can happily slide along the strip his entire life, believing himself to exist in an infinite universe. Why not? After all, he can perceive neither a beginning nor a terminus in his world. From our three-dimensional perspective, however, notice what occurs after he has completed one revolution around his universe, if we assume the "strip" to be transparent (see figure 7). After a trip around the strip, he is now the mirror image of himself and requires another revolution to return to his old form. One might contend that he is actually on the other side of the strip. But two-dimensional space has no "other side." The "strip" has created Zahir's enantiomorphic self through the *twist in three-dimensional space*. The same can be said of a three-dimensional

Chapter Four object turned inside out. Consider a two-dimensional drawing that is ordinarily perceived to be three-dimensional—the Necker cube—when placed before a mirror (figure 8).

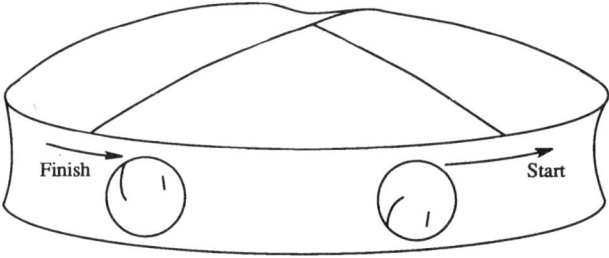

Figure 7

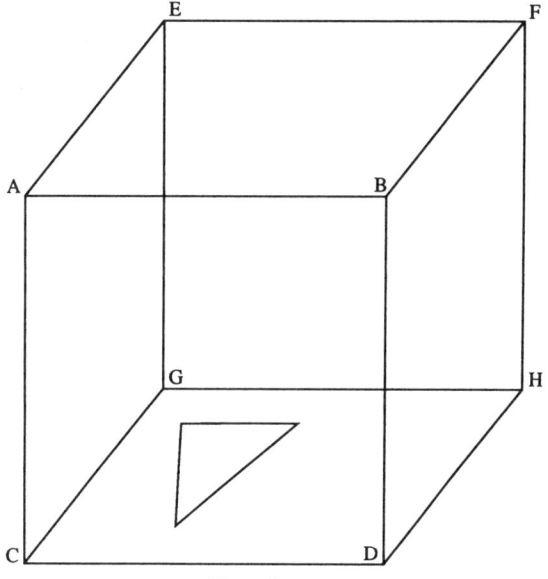

Figure 8

On the surface of the page, ABCD is on the *left side* and projected downward, whether toward you or away from you. In its mirror image, in contrast, ABCD is on the *right side* and projected downward, either toward you or away from you. Notice also that when ABCD is perceived to be the face of the cube, the triangle is on the "inside" with the apex pointing outward. On the other hand, if EFGH is the face, the triangle is "outside," and the apex points away. In other words, the triangle has been "twisted" in space, as was Zahir on the Möbius strip. The implication is that, just as the flatlander must be "rotated" through

three-dimensional space in order to be reversed, so the triangle must be "rotated," by means of an infinite numer of infinitesimals, through four-dimensional space in order to be reversed. And, just as the Möbius strip can be represented exclusively within a two-dimensional world, but with an intersection or "warp" in three-dimensional space, so the Necker cube can be conceived within a three-dimensional space, and its enantiomorphic transformation entails a four-dimensional temporal-spatial topology. Interestingly, a formulation analogous to this leads H. A. C. Dobbs (1972) to conclude that: (1) the transformation in the perspectival reversal of the Necker cube is logically impossible in three-dimensional space; (2) no spatial degree of freedom is evident in this transformation except height, width, and depth—that is, three-dimensional space; and (3) therefore the extra dimension must consist of a "mathematically imaginary time dimension," a four-dimensional manifold.

Now, for practical reasons, I have used the popular Necker cube to exemplify the necessity of a higher dimension for enantiomorphic transformations. Nonetheless, the same conclusions follow from other three-dimensional images, such as the hexagonal configurations making up the Library. By means of a mirror-image inversion, the same flip-flop effect of one of these images would equally require an extra degree of freedom, not spatial but rather a "mathematically imaginary time dimension." To repeat, the steps a given inhabitant of the Library takes, from birth to death, "trace an inconceivable figure in time" (*OI*, 128). The Library's mirrors, creating an infinite series of hexagonal enantiomorphs, hold more relevance to the "block" universe than initially meets the eye.

The books housed in the Library represent the second sphere of existence—the informational. These books are written by means of twenty-five symbols: the lowercase letters of the alphabet (excluding h, k, w, and x), commas, periods, and spaces. Each book contains 410 pages, each page 40 lines with 80 symbols each. Some claim that there is no totally nonsensical book in the Library: given the infinite number of civilizations, languages, and codes from past, present, and future, all combinations become intelligible at some time and place—though it is difficult to comprehend how a book containing the series *xyxyxyxy* . . . would not be nonsensical. Rucker (1983, 130) calculates that the number of symbols in each book in the Library is 410 pages × 40 lines × 80 symbols = 1,312,000. And since there are 25 different symbols, there are $25^{1,312,000}$ books, which is

approximately $10^{2,000,000}$! Lasswitz (in Fadiman 1958, 237–47), who presumably inspired Borges to write his story, obtains the same figure for his Universal Library, and to illustrate the magnitude of this number he estimates that a shelf $10^{1,999,982}$ light years long would be needed to hold them. This number of light years is so monstrously large that for practical purposes, it is not substantially smaller than the total number of books on the shelf! Understandably, for any given inhabitant within the Library, there is more than a little confusion regarding whether this overwhelming Babel (babble?) is finite or infinite.

Indeed, the Library is even more overwhelming than is our universe. The expanding universe is presently conceived to be homogeneous, i.e., it appears similar in all directions. Hence any conceivable center can only be both everywhere and nowhere, a concept originating with the Greeks and developed by Nicholas of Cusa (Borges's theme in "Pascal's Sphere"). Just as the universe consists chiefly of "empty" space, so the vast majority of the Library's books are unintelligible for a given researcher. But in a contest between our universe and the Library to boggle our minuscule minds, the Library wins hands down. If all matter in the universe were smoothed out uniformly, the density would be a mere 2×10^{-29} grams/cm^3. Under these conditions, assuming that we were alpha particles whizzing through the universe at a speed approaching the velocity of light, our chance of colliding with another particle of matter during a given second would be next to nil. Yet, upon considering the computed number of books in the Library, it becomes apparent that a given reader's chance of picking up a book with an intelligible first page is even more greatly reduced! Rather than the Library as metaphor of the universe, the inverse could be deemed more appropriately the case—the universe as a condensed metaphor of the Library.

This reintroduces the unsolved, and perhaps unsolvable, question, Are the books strewn throughout the Library at random or are they catalogued in some sort of order? Just as the narrator hardly gives us any answer, neither have modern mathematicians succeeded in giving us a precise definition of randomness, as was pointed out above. Randomness, of course, bears on the notion of infinity. It also presents difficulties for contemporary science, which is dependent upon probability theory based on the randomness of atomic events. If these events are indeed random, then there are limitless possibilities, but some are more probable than others, and the task is to discover the probability of a particular event's occurrence. The users of the

Library, in contrast, appear unable to manipulate their data; they cannot decipher the Library's system, for they are within the system, like the observer within his light cone in the "block" universe. In other words, they are apparently radical empiricists, or at best they are supernominalists, like Funes the Memorious, in contrast to the hypothetico-deductive physicist, and hence they possess no tools with which to construct a "probability theory."

But actually, there can be virtually unlimited types of order to the books in Borges's Library. For example, if you go to the nearest library, in minutes you can locate *Don Quixote,* for you are familiar with the system. However, suppose that, by some quirk, instead of the Library of Congress system, the books are arranged on shelves according to the color of their bindings, or by their type of print, their height, the number of words in the title, the number of times "and," "et," "y," etc., are used, and so on. There would be an order of sorts in each case, but how could you discover it? Now the difficulty of finding *Don Quixote* is magnified manyfold. Or, suppose universities to be fractured into departments with research specialists on books from particular presses. Professors in the Department of the University Press of Tlön would pass their time writing articles on the results of their research on those, and only those, books published by that particular press, all of which are found in a particular section of their library. If these professors were to be confronted with a library using the Library of Congress system, it would be a nightmarish, labyrinthine experience. The books would certainly appear heaped on the shelves at random. Though randomness and type of order depend in great part upon the perspective from which they are viewed, it has become evident, nonetheless, why any order to the bewildering maze of ciphers in the Library remains beyond the narrator's comprehension (Spencer-Brown 1957).

The Library's inhabitants constitute the third sphere of existence. Since this orb is organic in contrast to the Library's inorganic physical structure, I shall shift focus from the universe as Minkowski "block" to the statistically governed universe of the second law of thermodynamics. This entails the entropy principle, which, every physicist will tell us, is a subjective, anthropomorphic formulation—most appropriate when considering the human sphere of existence. Organic development of an individual organism is a complex process that can never reach complete, static, crystal-like symmetry. In other words,

Chapter Four

this process entails a decrease of asymmetry, like the inorganic process, but whereas the latter can ideally achieve completion, the former cannot. The chief purpose of the inhabitants of the Library is to discover order in the books; or, better, to invent order. The epitome of this desire is the revelation, supposedly received by some mystics, that somewhere there was a circular chamber with a circular book, the spine of which rests against the wall of the chamber. This is the perfectly symmetrical library, cyclical and unbounded, like, according to various testimonies in Borges's "Pascal's Sphere," God himself. It is also an adequate vision of the total Library. But, from within the apparently random collection of letters, words, and books, this image is potentially anything and everything: "No one can articulate a syllable which is not filled with tenderness and fear, which is not, in one of these languages, the powerful name of a god. To speak is to fall into tautology" (*L*, 57).

In reality, the most improbable state of the Library *is* orderliness. When a system is orderly, it is low in entropy and rich in structure; it bears a relatively higher degree of information than when it is disorderly, with maximum entropy. One way of construing this elusive principle is that the higher the entropy and the greater the disorder, the more numerous are the possible states of affairs. That is, with a large number of possible combinations, entropy is high; with fewer combinations and hence increased predictability, entropy is minimized. Abolishing all grammar rules of the English language would increase entropy; with the addition of more restrictive rules, lower entropy would ensue, but at the expense of a degree of freedom. Since high entropy indicates lack of possible information and greater uncertainty, the Library is obviously in a random or near-random state. The demon (or God) who arranged the letters in the books is the prototype of the proverbial army of monkeys pecking away on typewriters. Another demon, that of Maxwell, it might be assumed, could possibly be a Super-librarian capable of putting the disarray of books in order; only he could extract information from each and every book. But apparently no such omniscient demon exists.

Interestingly, Spengler, who singled out entropy as the greatest contributor to the downfall of classical science due to its statistical character, wrote:

> Statistics belongs, like chronology, to the domain of the organic, to fluctuating life, to Destiny and Incident and not to the world

of laws and timeless causality. . . . In the "classical" mechanics of Galileo and Newton there would have been no room for them. And if, now suddenly the contents of that field are supposed to be understood and understandable only statistically and under the aspect of Probability, . . . what does it mean? It means that the object of understanding is ourselves. (Spengler 1932, 421)

The Universe as Library

Living organisms processing information from a statistically determinable world partly offset the unidirectional movement toward entropic decay. Birds build nests, beavers build dams, ants organize themselves into structured communities, humans both build and destroy. In the Library, unfortunately, almost pure "noise" prevails, but not quite; barely enough information is received, after one has toiled for years, to nurture a feeble ray of hope. *Inquisitors* are sent out on tireless journeys in search of the origin and meaning of the Library. A blasphemous sect suggests that the search should cease and "that all men should juggle letters and symbols until they constructed, by an improbable gift of chance," the canonical books (*L*, 56–57). Others take refuge in the latrines "with some metal discs in a forbidden dice cup and feebly mimic the divine disorder" (*L*, 56).[10] But this hope was followed by depression. The narrator implores to God: "[B]ut for one instant, in one being. Let Your enormous Library be justified" (*L*, 57). Alas, the order remains in the final analysis unknowable; despair creeps in, decay spreads:

> Epidemics, heretical conflicts, peregrinations which inevitably degenerate into banditry, have decimated the population. I believe I have mentioned the suicides, more and more frequent with the years. Perhaps my old age and fearfulness deceive me, but I suspect that the human species—the unique species—is about to be extinguished, but the Library will endure. (*L*, 58)

Significantly, as an aside, the astronomer David Layzer (1975) recently put forth the radical hypothesis that even if a Laplacean Superobserver could exist, total information about the universe at the molecular scale would still be impossible. He concludes:

> The order is unknowable even in principle. Imagine an unbounded stack of playing cards, topless and bottomless, deck piled on deck without limit. Information about the order of the cards in one section of the stack is of no help, because any given

Chapter Four

sequence is repeated an infinite number of times elsewhere, in the same way that patterns of stars and galaxies are repeated throughout the universe. It is meaningless to say that you are at such and such a place in the stack, even when you have full information about the order of the cards and that place. You still don't know whereabouts you are in the stack, any more than the typical observer knows whereabouts he is in the cosmos. The Cosmological Principle says he doesn't know and he can't know. (in Campbell 1982, 89)

This stack of cards is indeed a worthy counterpart to Borges's Library, whether conceived as geometrical structure, a monstrous set of books, or, from the perspective of the Library's researchers, a statistical conglomerate, the meaninglessness of which is practically absolute.

Chapter Five **Chronos in Chains**

Another issue that has surfaced in the present century is the elusive nature of time, or better, its contrary, timelessness. Two diametrically opposed views pervade the massive body of literature regarding time: (1) that "reality" is pure change (becoming), and (2) that it is pure stasis (being). The apparent mutual exclusiveness delineating these concepts coincides with other time-related concepts. *Necessity* and *contingency* distinguish between events that must occur and possibilities that may or may not come about; the *one*, which is identical with itself, is diametrically opposed to the *many*, which calls for permutability; *infinity*, which is eternal sameness, contradicts *finiteness*, which entails foregrounding and backgrounding. One class of categories appears timeless, while the other is temporal. Neither of these modes of thinking about time is comprehended by the other, nor can one be reduced to the other. They are at most *complementary*, each supplementing the other, with neither giving full account of what there is.

Borges's well-known refutation of time is not unique. Given its aesthetic qualities, it has few rivals, but it enjoys a number of metaphysical counterparts, some of which will be

mentioned. After a review of Borges's essay, I offer a general discussion of time and timelessness from the perspective of the Einsteinian cosmology. The perennial philosophy—mystical experience—which has been of interest to numerous secular scholars since Hinton, will be commented upon briefly. Then, a discussion of "space-time singularities," or "black holes," where time is dramatically halted altogether, and their remarkable affinity with Borges's Aleph, will conclude this chapter.

According to Jorge Luis Borges it is not reading but rereading that matters. If this is so, what is the presumptive fate of books in a society which identifies time with continual change?
—*J. T. Fraser*

Time's Eternal Struggle

1 Borges, regarding time, once confessed:

I tend to be always thinking of time, not of space. When I hear the words "time" and "space" used together, I feel as Nietzsche felt when he heard people talking about Goethe and Schiller—a kind of blasphemy. I think that the central riddle, the central problem of metaphysics—let us call it thinking—is time, not space. Space is one of the many things to be found inside of time—as you find, for example, color or shapes or sizes or feelings. (Christ 1972, 400–401)

Borges's obsession with and perpetual awareness of time becomes forcefully evident in his comment on the title of "New Refutation of Time." He is not unaware, he reveals, that the title is "an example of the monster which logicians have called *contradictio in adjecto*" (*OI,* 172), for to declare a refutation of time to be either new or old is to attribute to it a predicate of temporality, thus retaining the very concept the subject is supposedly to annihilate. Nonetheless, Borges decides to let the title stand as a "subtle joke," demonstrating that he does not exaggerate the importance of such word games. He then remarks on the impossibility of freeing ourselves from temporality altogether, for "our language is so saturated and animated with time that it is very possible that not one line in this book does not somehow demand or invoke it" (*OI,* 172).

Borges, who has dedicated a lifetime to literature and occasionally to "metaphysical perplexities," has sensed a denial or refutation of time, "which I myself disbelieve, but which comes to visit me at night and in the weary dawns with the illusory force of an axiom" (*OI,* 172). Yet his writings tend to belie this disbelief in the passage of time. Many of his texts are narrated as if they existed in the perpetual now: a timeless, static present, a "block."

For instance, we are told in "The Garden of Forking Paths" that "everything happens to a man precisely, precisely *now.* Centuries of centuries and only in the present do things happen" (*L,* 20). The narrator reflects in "The Immortal" that "to be immortal is commonplace; except for men, all creatures are immortal, for they are ignorant of death" (*L,* 114). In "The South," Juan Dahlmann, traveling to a remote part of Argentina for a period of convalescence following a near-fatal accident, also travels back in time, or perhaps into the timeless. He enters a café where there is an enormous black cat, which he caresses, but he senses that "this contact was an illusion and that the two beings, man and cat, were as good as separated by a glass, for man lives in time, in succession, while the magical animal lives in the present, in the eternity of the instant" (*F,* 169–70). Emma Zunz, in the story by the same name, decides to take revenge against Lowenthal, who had caused her father to be accused and exiled and finally to commit suicide. She sells herself as a prostitute to a sailor, then goes to Lowenthal, kills him, and declares that he had violated her and she was forced to kill him in self-defense. The events of her prostitution do not happen *to* her as part of her lived and experienced succession of events. They remain "outside of time, either because the immediate past is as if disconnected from the future, or because the parts which form these events do not seem to be consecutive" (*L,* 134). This is analogous to Hladik, whose instant of now, when time stops, is all there is; it is, for the duration of one year, eternal. And finally, in "The Theologians" the narrator discloses his dilemma: "The end of this story can only be related in metaphors since it takes place in the kingdom of heaven, where there is no time" (*L,* 126). This is necessarily the case, since all language "is of a successive nature; it is not an effective tool for reasoning the eternal, the intemporal" (*OI,* 178).[1]

Borges begins "New Refutation of Time" with another apology. Published in the eighteenth century, his essay would "endure in the bibliographies of Hume and perhaps would have

been mentioned by Huxley or Kemp Smith" (*OI,* 171). Published after Bergson, however, it is "the anachronous *reductio ad absurdum* of an obsolete system or, what is worse, the feeble machination of an Argentine adrift on the sea of metaphysics" (*OI,* 171). Despite this, Borges considers his concept of time to be the ultimate consequence of the combined doctrines of Berkeley and Hume. In a nutshell, Berkeley, Borges reveals, denies the existence of an external reality independent of our perception of it, while he retains the notion of a perceiving subject. Hume discounts the existence of a perceiving subject; the subject is merely a bundle of sensations. That is to say, we cannot speak of the form and color of the moon; they *are* the moon. Nor can we speak of the mind's perceptions; the mind is *nothing more than* a series of perceptions.

Borges then evokes a scene in which Huckleberry Finn awakens while on his raft during one of his numerous nights on the Mississippi. He "opens his eyes lazily. He sees a vague number of stars, he sees an indistinct streak of trees; then he sinks into an immemorial sleep that envelops him like murky water" (*OI,* 176). We are reminded that the "metaphysics of idealism declare that it is risky and futile to add a material substance (the object) and a spiritual substance (the subject)" to the equivalent of Huckleberry Finn's mindless perceptions. Borges, on the other hand, maintains that "it is no less illogical to think that they are terms of a series whose beginning is as inconceivable as its end" (*OI,* 176). Though this series is admittedly "illogical," it is asserted, nonetheless, that time is indeed a series rather than continuous. Combining Berkeley with Hume, both external reality and the subject disappear. Perception occurs solely in the present, and if there can be a single instant of repetition, that is, of two identical moments, then that will be evidence enough, Borges asserts, to deny time altogether.

As a testimonial for such a denial, Borges subsequently presents his "feeling in death" experience at a street corner in Barracas that is identical to an experience at the same spot some thirty years earlier. Then he concludes: "If we can perceive that identity, time is a delusion: the indifference and inseparability of one moment of time's apparent yesterday and another of its apparent today are enough to disintegrate it" (*OI,* 180).

In her rigorous analysis of "New Refutation," N. Katherine Hayles (1984, 162–63) demonstrates how Borges, almost imperceptibly, enacts a shift in his discourse from "continuity" to "series" when speaking of time's flow. First referring to tem-

poral continuity, he then switches to time as a discrete series, which introduces the equivalent of a Zenoesque argument and hence a paradox. Time as serial, and its relevance to Zeno, also implies the problem of change. In his essay, Borges recounts the dream of Chuang Tzu that he was a butterfly, and while dreaming, he *was* a butterfly; that is, he had *changed into* a butterfly. But upon awakening, he could not decide whether he was (i.e., had reverted into) Chuang Tzu who dreamed he was a butterfly or a butterfly that dreamed it was (i.e., had *changed into*) Chuang Tzu. Colin Butler (1973, 157) points out that Chuang Tzu must have been either one or the other, and while his identity was indeterminable, the fact *that* this is so "can only be the consequence of an act of recollection, with its implication of change, and therefore of succession, which his experience was intended to controvert."

However, Butler's notion of change is not so clearly implied if we follow F. H. Bradley (1897, 39ff)—who is cited by Borges in "New Refutation" and elsewhere. If, says Bradley, an object X changes, then it is either in time or out of time. If it is out of time, then we have on the one hand the timeless X and on the other the various serial recollected states, $X_1, X_2, X_3, \ldots X_n$, which lie in time. The question then arises, What is the relation of X to its serial states? If there is no relation, then in no sense does X change—and hence Chuang Tzu could neither have become the butterfly nor the butterfly Chuang Tzu. But if there is a relation, then how can a timeless object be intelligibly related to the succession in time? If we cannot explain this, the removal of X from time is logically impossible. Assume, then, that X is within serial time. The question then arises, Does it or does it not have duration? If it has duration, then it falls apart into its disconnected successive states, and we must bow to Zeno, for logically speaking, there can be no process of change from one state to another. If, on the other hand, it has no duration, then, says Bradley, there still can be no change, for change implies a condition before and another condition after, and in a simple "now" this becomes impossible. Change, therefore, is impossible.

J. M. E. McTaggert (1927, ch. 33) argues with Bradley—and, indirectly, with Borges—that time is unreal. There are two explicit sequences of relations in time: (1) a transition from past to present (which is not infinitely divisible), and (2) the continuum of earlier to later (which is infinitely divisible). The first is self-contradictory, says McTaggert. Since it assigns three incompatible determinants to a single event, it gives no

understanding either of change or time, and hence the two sequences (past-present-future and earlier-later) cannot be made one. Hence a state of timelessness must prevail.

Borges, from a slightly different framework, assures us that to deny time is actually two denials: "the denial of the succession of the terms of a series, the denial of the synchronism of the terms of two series" (*OI*, 185). If time consists of a succession of "nows," or terms, these "nows" prove to have a contradictory character. Each "now" must admit of relations with other "nows," and each "now" must be either divisible or indivisible. But it is not indivisible, for if so, it would have neither a beginning connecting it to the past nor an end connecting it to the future; nor could it have a middle, for it there is no beginning or end, there can be no middle. Nor can it be divisible, for if so, "it would consist of a part that was and another part that is not. *Ergo,* the present does not exist, and since the past and future do not exist either, times does not exist" (*OI*, 185–86).

Finally, after construction of his dialectic of Berkeley and Hume, Borges arrives at Schopenhauer's statement according to which time is "like a rotating circle: the descending arc is the past, the ascending one is the future: above there is the indivisible point that touches the tangent and is the now" (quoted in *OI*, 186). The cycle is now completed; Borges has reverted back from serial time to a continuous notion of time. He also relates Schopenhauer's metaphor to another similar one from a fifth-century Buddhist tract stating that "the life of a being has the [continuous] duration of an idea. As a carriage wheel touches the ground in only one place when it turns, life lasts as long as a single idea" (*OI*, 186). Borges now concludes his essay:

> And yet, and yet—To deny temporal succession, to deny the ego, to deny the astronomical universe, are apparent desperations and secret assuagements. Our destiny . . . is not horrible because of its unreality; it is horrible because it is irreversible and ironbound. Time is the substance I am made of. Time is a river that carries me away, but I am the river; it is a tiger that mangles me, but I am the tiger; it is a fire that consumes me, but I am the fire. The world, alas, is real; I, alas, am Borges. (*OI*, 186–87)

Some critics allude to Borges's own admission of failure to refute time effectively and to his confession of disbelief in his own argument as a defeat. For example, Butler (1973, 154)

remarks that "New Refutation" is "in essence the consciously futile attempt by Borges to subordinate his knowledge of transitoriness to the hope of eternity by affirming the latter at the expense of the former."[2] On the other hand, Albert Bagby (1965, 102–3) places stock in the Argentine essayist:

> Borges indicates to us—with reason—*that time in the form of succession becomes easily disrupted at the exact moment in which any element of this succession is repeated.* Thus Borges is correct in pointing out that if outside of each awareness (actual or conjectural) matter does not exist; and if also outside of each mental awareness the spirit does not exist, neither will time exist outside of each present moment.

Yet the fact remains that Borges engages in a fusion of continuity and discontinuity, which culminates in an argument ad absurdum. And, what is more disconcerting, the same type of argument appears elsewhere. For instance, time as series recalls Borges's essay on Dunne, for whom "real time is the unattainable final boundary of an infinite series" (*OI,* 20). But Dunne's conception of time is invalid, since increments along the series must be eventually repeated, and a single repeated term, Borges declares, disrupts the series—"time" now being considered as a static collection of disconnected increments. In "The Secret Miracle," Borges refers to the second book Hladik is writing, which follows Bradley's argument that all the events in the universe compose a temporal series and man's possible experiences are therefore finite. Then, reiterating the argument in the "New Refutation," Borges asserts that a single repetition suffices to prove that time is a fallacy. But unfortunately, he confesses, "the arguments that demonstrate this fallacy are equally fallacious" (*L,* 90).

While it is true that Borges's fusion of continuity with a discrete series is contradictory, it is precisely what Zeno accomplishes. Similar arguments have also been used by the most accomplished of metaphysicians. Furthermore, as we shall observe below, Borges's essay is not as outdated as he would lead us to believe; it is in line with some of the most rigorous of contemporary thought. For example, Whitehead (1929, 69) remarks on Achilles and the Tortoise:

> Consider . . . an act of becoming during one second. The act is divisible into two acts, one during the earlier half of the second, the other during the later half of the second. Thus that which

becomes during the whole second presupposes that which becomes during the first half-second. Analogously, that which becomes during the first half-second, and so on indefinitely. Thus if we consider the process of becoming up to the beginning of the second in question, and ask what then becomes no answer can be given. For, whatever creature we indicate presupposes an earlier creature which became after the beginning of the second and antecedently to the earlier creature. Therefore there is nothing which becomes, so as to effect a transition into the second in question.

For Whitehead, every event must have a predecessor and a successor, but the event in which Achilles overtakes his competitor can have no predecessor if each of the infinite number of steps comprises an event. The point is that our experiential understanding of time (and of becoming in the sense of Whitehead) is that of something divisible and divided, and change is understood in terms of two conditions: (1) an infinitely divisible continuum and (2) a succession of events (a series). But the two are mutually exclusive. The problem is that even though time may be divisible, events are not. This relation between continuity and a discrete series appears to have no ready logical solution. Yet, Whitehead claims, it *is* resolved through our concrete experience of becoming; or better, experience dissolves the problem.

Whitehead's notion of an infinitely divisible continuum originates in mathematics. For example, Dantzig (1930, 175–78) demonstrates how what are called "Dedekind cuts" in a mathematical continuum (i.e., a line) are related to time. Richard Dedekind's axiom states: "If all points of a straight line fall into two classes, such that every point of the first class lies to the left of any point of the second class, then there exists one and only one point which produces this division of all points into two classes, thus severing the straight line into portions" (Dantzig 1930, 177). It readily becomes obvious that this severing is a skillful paraphrase of the fundamental characteristic we ordinarily attribute to time, where the "now" separates past from future, both of which are mutually exclusive, and both of which comprise Borges's notion of timelessness. Paradoxical though it may seem, the present "now" is irrational in Dedekind's sense, for while it acts as a boundary, it is neither past nor future; it is a mathematical fiction. Neither Cantor nor Dedekind were able to emancipate continuity from the intuition of time. The peren-

nial conflict between the continuous and the scientific concept of discrete number series ended in a victory for the latter, a victory consummated with the invention of analytic geometry. And by fiat, as it were, space (that is, the line) became a mere bearer of numbers with the "Cantor-Dedekind axiom," which is: "It is possible to assign to any point along a line a unique real number, and conversely, any real number can be represented in a unique manner by a point on a line" (Dantzig 1930, 177–78). And all this, by exercising an abstractive act of mind!

Our "intuition" of time, in the sense of Whitehead, is perhaps no more nor less a mental act than the axiom of the line and real numbers. Perhaps we may find here a more adequate vindication of Borges's "New Refutation" which, in spirit though not entirely in form or content, can render Borges's essay contemporary with the concept of time in physics. As discussed above, in the Minkowski "block" universe there is no before or after; everything exists in the eternal "now"; becoming is mind-dependent. Olivier Costa de Beauregard (1981, 430) states that relativity is

> a theory, in which everything is "written" and where change is only relative to the perceptual mode of living beings. Humans and other living creatures, for reasons which one can try to explain are compelled to explore little by little the content of the fourth dimension, as each one traverses, without stopping or turning back, a time-like trajectory in space-time.

Our everyday experience leads us to believe in a world of three-dimensional space in which things change with the passage of time. The past is gone, the future has not yet arrived, and only the ever-moving present is "real." To regard the universe as a space-time continuum, the "block," is of course in a manner of speaking to stand outside history. This, we might suppose, is the essence of Borges's "mystical" experience on the street corner in Barracas. It should come as no surprise that this is also a universal feature of the *philosophica perennia:* what we experience as the succession of time is an abstraction rather than "reality," and "the real state of the universe is eternal or timeless" (Watt 1954, 29). This presents a problem. If what we experience is no more than an abstraction, then how can we get at the real "real"? Our commonsensical view of things fails us; the "real world" as we know it fails us also. So where, if there be any, is the handle we can grasp onto in order ultimately to know "reality"? Though in all probability

Chapter Five

no answer can be forthcoming to this question, the mystic, philosopher, and mathematician P. D. Ouspensky (1922, 106) might have revealed as lucidly as anyone how things might be conceived and perceived outside our normal perception:

> Let us imagine some object, say *a book,* outside of time and space. What will this last mean? Were we to take the book out of time and space it would mean that *all books* which have existed, exist now, and will exist, *exist together,* i.e., occupy one and the same place and exist simultaneously, forming as it were *one book* which includes within itself the properties, characteristics and peculiarities of all books possible in the world. When we say simply, *a book,* we have in mind *something* possessing the common characteristic of all books—this is *a concept.* But that *book* about which we are talking now, possesses not only these common characteristics but the individual characteristics of all separate books. . . .
>
> What is "man" out of space and time? He is all humanity, man as the "species"—*Homo Sapiens,* but at the same time possessing the characteristics, peculiarities and individual earmarks of *all* separate men.

Note the remarkable similarity between this quote and a host of Borges's comments on, for example, the Aleph, a tiger being simultaneously all tigers, a man all men, the fabled infinite Book of all books in the Library, Dahlmann's journey south and back in time, and so on, where all is *there,* timelessly, in the only legitimate "reality" which remains outside our perception of things.

With respect to a state of timelessness considered purely as a "mental act," interestingly, Gödel contributed a relatively little-known paper to the "Library of Living Philosophers" volume on Einstein, in which he notes that one of the most intriguing aspects of relativity theory "consists in the fact that it gave new and surprising insights into the nature of time, of that mysterious and seemingly self-contradictory being which, on the other hand, seems to form the basis of the world's and our own existence" (Gödel 1949, 557).[3] Gödel explains that a remarkable property of time, derived from relativity, is non-simultaneity and the indeterminacy of the succession of events. The assertion that A occurred before B loses its objective meaning, for from the vantage point of another observer within the "block," B could well have occurred before A. The consequence ensues from this strange state of affairs that there now

seems to be "unequivocal proof" for the view of Parmenides, Kant, and the modern idealists, who denied the objectivity of change and time.[4] In fact,

> [c]hange becomes possible only through the lapse of time. The existence of an objective lapse of time, however, means (or at least is equivalent to the fact) that reality consists of an infinity of layers of "now" which come into existence successively. But, if simultaneity is something relative . . . reality cannot be split up into such layers in an objectively determined way. Each observer has his own set of "nows," and none of these various systems of layers can claim the prerogative of representing the objective lapse of time. (Gödel 1949, 558)

Given the relativity of simultaneity, Gödel continues, since space-time cannot be sliced and stacked into a unique set of "nows," the "now" of each observer is relative to all other "nows," and past and future are indeterminately relative. Gödel then uses an argument comparable to that of Borges: thoughts have a reality external to us, since more than one person can have the same thought within the timeless "block," hence regarding the "block" in its entirety, that same thought occurs in absolute simultaneity, i.e., irrespective of any given subject's "time line." Then Gödel asks us to imagine a rotating universe in which the local times of a certain number of observers cannot constitute a single world time, i.e., some of the "world lines" would appear to some observers to be simultaneous, but for others some would be future and others past, and for still other observers the order would be inverted. Obviously these observers' perception of change and hence of time would be subjective; change is not a reality but a frame of reference, a perspectival vantage point.

For example, reconsider Butler's critique of Borges's use of Chuang Tzu's inability to determine whether he dreamed a butterfly or vice versa. According to Butler, this identity crisis could only be predicated on an act of recollection, which implies both change and succession. However, following Gödel—and in light of Bradley and McTaggert—recollection has nothing to do with past series of events, nor do future expectations have to do with future series. Recollection itself is within the monolithic "block." Hence, just as two or more persons can have the same thought within this timeless sphere, so they can have the same recollections or expectations, whether in the form of thought,

Chapter Five dream, hallucination, or whatever—recall Borges's essay on Coleridge.

In sum, the point of Gödel's captivating paper is that linear experience is not given; it is a mental construct. The flow of time, we ordinarily believe, is invariably from past to future, but according to Gödel the contrary is no less logical. This state of affairs implies an absurdity: time travel in the past to a place where one has already lived such that one could alter one's own future. For practical purposes, however, and as Gödel himself confesses, the velocities required for such travel are far beyond what might conceivably become possible in the remote future.

Significantly, in light of Gödel's thesis, Borges, after remarking also that the passage of time from future to past is no less conceivable than the inverse, refers to one of Bradley's hypotheses, which proposes that we

> exclude the future, which is a mere fabrication of our hopes and . . . reduce the "actual" to the agony of the present moment disintegrating into the past. . . . Bradley denies the future; one of the philosophical schools of India denies the present, for it is considered ungraspable. *The orange is about to fall from the branch, or it's already on the ground,* those simplifiers of reason affirm. *Nobody sees it fall.* (*HE*, 12–13)

The orange exists in two distinct synchronic frameworks; "time," our illusion of time, is of no consequence. We have here once again the Buddhist arrow paradox. I see the arrow poised in the strained bow, *ready for flight;* someone else sees it embedded in the bull's-eye of the target, *its destination having been completed.* There is no before or after, however: the Buddha sees it all from the "block" perspective, sempiternally.

After this deliberation on time, or better, timelessness, let us now shift to space—that is to say, "spacelessness," which will reveal the ultimate extension of Einsteinian physics.

> *Beginning with the big bang, the universe expands and cools. After eons of dynamic development it gives rise to observership. Acts of observer-participancy—via the mechanism of a delayed-choice experiment—in turn gave tangible "reality" to the universe not only now but back to the beginning.*
> —John Wheeler

Chronos in Chains

2 Rucker (1977, 120) once speculated that Borges's Aleph might possibly be conceived of as a description of what a "space-time singularity" might look like. I intend in this section to carry this idea further. Borges (*A*, 263–64) comments, regarding his tale of the Aleph:

Singularities and Other Strange Phenomena

> What eternity is to time, the Aleph is to space. In eternity, all time—past, present, and future—coexists simultaneously. In the Aleph, the sum total of the spatial universe is to be found in a tiny shining sphere barely over an inch across. . . . [W]ere such an object to exist it would not only be the most famous thing in the world but would renew our whole conception of time, astronomy, mathematics, and space.

Borges's comparison of the Aleph to eternity is apt. Eternity is timeless, yet it "contains" the whole of time, past, present, and future. The Aleph is, for practical purposes, virtually spaceless. And it "contains" all things in the universe, including their temporal sequences, in a static, timeless instant.

Borges undoubtedly was not aware in 1970, when he published his "Commentaries" to *The Aleph and Other Stories,* how true his words would ring. Throughout his texts Borges enumerates radically divergent elements, the Aleph being the most noteworthy, that mirror or refer to some static totality that is indescribable in natural language. Cándido Pérez Gallego (1966, 119) compares "The Aleph" to H. G. Well's "The Crystal Egg," which enables a grasp of eternity, and he relates it to the tradition of the British Gothic novel. Be that as it may, the Aleph, I believe, more fittingly enjoys a contemporary counterpart in the concept of a space-time singularity, especially the so-called "naked" singularity.

Chapter Five

Following from Einsteinian relativity, it is now conceived that space-time singularities can theoretically be produced when a star collapses. Ordinarily stars do not collapse due to their own gravitational attraction because this attraction is offset by their tendency, as hot masses of gas, to expand. As a star cools, however, it can contract and become more dense until finally the shrinkage causes it to remain at a fixed but tiny radius, more adequately described as a fixed point than a condensed "lump." To illustrate such a fixed point, space-time can be represented by a cone, the surface of which is concave (see figure 9; also Davies 1981, 68). During the creation of a singularity, space-time curves along its "time line," i.e., the surface of the cone, due to the collapsed star's tremendous gravity, until it is finally pinched into a point. At this stage the limits of time and space have been reached. An observer, a particle, or a ray of light traveling along the surface of the cone risks being sucked into this spaceless, timeless singularity, never to return. If this description is correct, however, an absurdity apparently ensues, for if the density of the star is increased without limit, it becomes, theoretically, infinitesimally small and infinitely dense. In other words, near the base of the cone the curvature is relatively slight, but on traveling from the base to the tip of the cone the curvature becomes infinite (Davies 1981, 92–129).

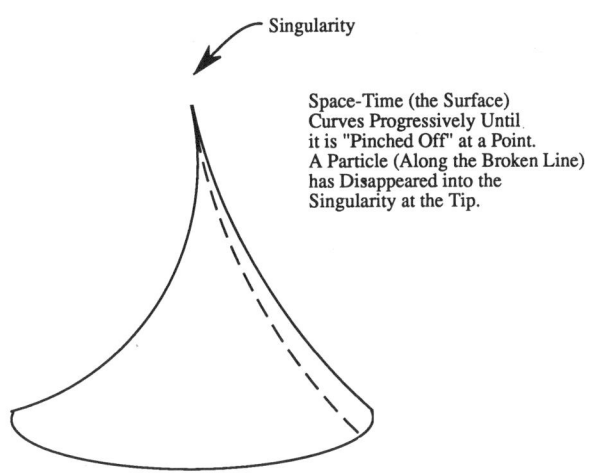

Figure 9

Here we suffer another confrontation with Zeno. Imagine space-time events to be labeled by numbers, with zero corresponding to the singularity—which is actually not an event, for at that point space-time has ceased. At what time can a first event after the singularity occur? After one-millionth of a second? One-trillionth? No matter how many times we divide the fraction, we shall never reach the smallest number, and hence the first event is logically impossible (Davies 1983, 18–19). Another way of putting this is that there are no fewer "instants" in the final increment following the space-time singularity at the apex of the cone than there are along all places seen from all angles along the surface of the cone. In this sense, the entire cone can be conceived as one gigantic (ineffable) "instant" of all points in space without any overlapping (Davies 1983, 23–27).

Ordinary singularities are the product of "black holes," whose gravitational force is of such magnitude that light is trapped within them. There are also special types of singularities, "naked" singularities, which are theoretically formed in the absence of black holes. Various hypotheses on the characteristics and possible formation of naked singularities exist—and, it must be admitted, some physicists disbelieve their existence altogether. In general, the difference between a black hole and a naked singularity—"unclothed" with a black hole—is that light can escape one but not the other (Davies 1981, ch. 6). In the formation of both, a star implodes, shrinking virtually to nothing. The black hole folds and convolutes light back into a surface, called the "event horizon," which "clothes" the singularity. Nothing can cross this horizon to the outer region, so it remains invisible to the observer (see figure 10; also Davies 1981, 110). In contrast, in the collapse of a star to a naked singularity, light rays are convoluted around the singularity, which is a charged, rotating hole unlike the nonrotating black hole, and after spiralling around it for a period of time, the rays can finally escape to the outer region, and hence theoretically they now become observable. One hypothesis has it that the time condensation for the observer gazing into a rapidly rotating and highly charged naked singularity affords him an extraordinarily dramatic moment, for the singularity is exposed to and exposes the entire universe. Like Borges's contemplation of the Aleph, one can, when gazing at a naked singularity, witness in a flash the total history of the universe compressed into a moment. This implies that the universe's

Chronos in Chains

Chapter Five   information is also compounded in time; in an instant an infinite quantity of light arrives.[5]

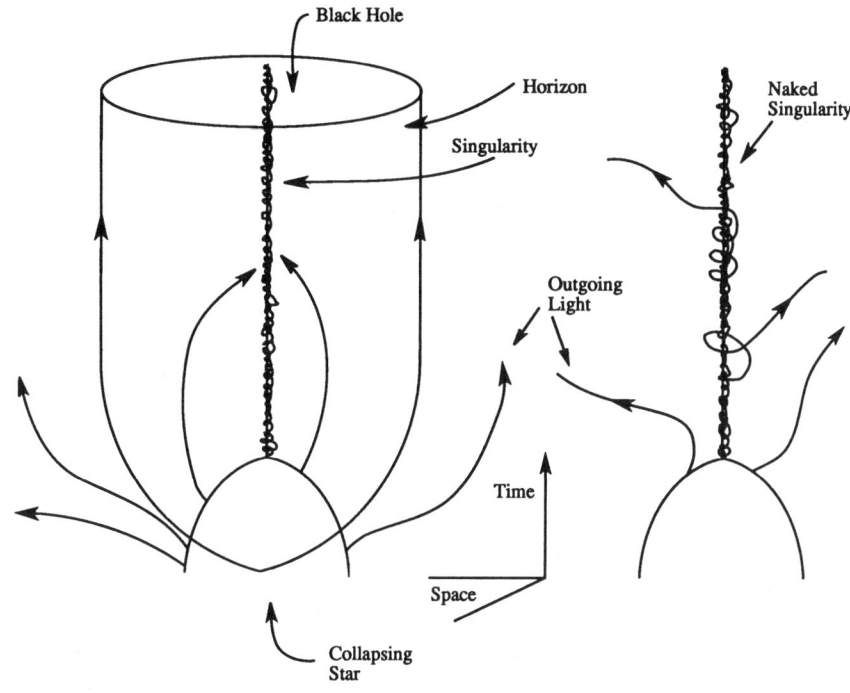

Figure 10

Significantly, Borges tells us that "all stars, all lamps, all sources of light" were in the Aleph, and that the light emanating from it was of "unbearable brilliance" when he saw it in that single "gigantic instant." There were

> millions of acts both delightful and awful; not one of them amazed me more than the fact that all of them occupied the same point in space, without overlapping or transparency. . . . [A]ll space was there, actual and undiminished. Each thing (a mirror's face, let us say) was infinite things, since I distinctly saw it from every angle of the universe. . . . I saw in a backyard of Soler Street the same tiles that thirty years before I'd seen in the entrance of a house in Fray Bentos; . . . I saw in a closet in Alkmaar a terrestrial globe between two mirrors that multiplied it endlessly; . . . I saw the delicate bone structure of a hand; I saw the survivors of a battle sending picture postcards; I saw in a showcase in Mirzapur a pack of Spanish playing cards; I saw the slanting shadows of ferns on a greenhouse floor; I saw tigers,

pistons, bison, tides, and armies; I saw all the ants on the planet; . . . I saw the circulation of my own dark blood; . . . I saw the Aleph from every point and angle, and in the Aleph I saw the earth and in the earth the Aleph and in the Aleph the earth; I saw my own face and my own bowels; . . . I felt infinite wonder, infinite pity. . . . I was afraid that not a single thing on earth would ever again surprise me; I was afraid I would never again be free of all I had seen. (*A*, 26–28)

But fortunately, after a few "sleepless nights," the narrator's experience passed into oblivion, as if it had all been a dream. Besides the Aleph's obvious allusions to infinite sets and the parallels with naked singularities, it is worthy of note that the bones of a hand, Borges's own bowels, a piston inside an engine, or articles inside a closet, all would be open to view from a four-dimensional perspective, just as in a comparable sense a three-dimensional spherelander could see inside the closet belonging to a two-dimensional flatlander or could watch the aspirin the flatlander took for a headache pass through his esophagus and enter his stomach.

Naked singularities may be the ultimate unknowable. At a singularity, matter can either enter or leave the ordinarily known physical world, and the consequence of such exchanges may well be completely beyond prediction. It seems that what will emanate from naked singularities boils down to two alternatives: either total chaos and the unstructured, or coherent, totally organized events (Davies 1983, 55–56). If a naked singularity were to form near the earth tomorrow, would the universe become a madhouse? An unruly mass of chaotic events? Would we all be destroyed? There are some prophets of doom who, with terrible foreshadowings, predict that a singularity "could destroy our orderly universe. It could be that only by construction of such a monstrosity would we ever learn what really does happen to matter at the final stage of collapse to a nonexistent point. If so, it would be a hard price to pay—the end of our own world—for discovering what such a finale really is" (John Taylor in Gribbin 1977, 174).

Whether tame or vicious, random or ordered, an "ignorance principle" prevails regarding the nature of singularities, and it may continue to do so indefinitely: a naked singularity, in the final analysis, should not be regarded as an object or a thing so much as a nonplace where all known laws are suspended (Davies 1983, 56–57). The ineffable Aleph could likewise be so regarded.

Chapter Five

There is another point of contact between Borges's Aleph and singularities: mystical insight. The legion of hypotheses and speculations—even a few crackpot ideas—on singularities and other such strange phenomena appear to belong more to religion or metaphysics than to science. They at times even border on mysticism—significantly, Borges (*A*, 25) refers to the Aleph as "the microcosm of the alchemists and Kabbalists." Indeed, as I have mentioned, attempts to discover analogies between the "new physics" and Eastern religions are notorious. The truth of the matter is that the two systems simply are not fraternal twins. Though certain parallels admittedly exist between contemporary physics and Oriental thought, the similarities between them are usually trivial when compared to the differences.[6] Moreover, physics neither proves nor disproves, nor does it support or refute, mysticism. Certain of the greatest twentieth-century scientists cultivated a mystical view of the universe, notably Einstein, Schrödinger, Eddington, and Jeans. Yet they categorically denied that their theoretical frameworks enjoy any determinable correlation with their mystical views. Jeans (1930, 135) summarizes his thoughts thus:

> The essential fact is simply that *all* the pictures which science now draws of nature, and which alone seem capable of according with observational fact, are *mathematical* pictures. . . . They are nothing more than pictures—fictions if you like, if by fiction you mean that science is not yet in contact with ultimate reality. Many would hold that, from the broad philosophical standpoint, the outstanding achievement of twentieth century physics is not the theory of relativity with its welding together of space and time, or the theory of quanta with its present apparent negation of the laws of causation, or the dissection of the atom with the resultant discovery that things are not what they seem; it is the general recognition that we are not yet in contact with ultimate reality. We are all imprisoned in our cave, with our backs to the light, and can only watch the shadows on the wall.

Here, Jeans outlines the remarkable distinction between classical physicists and the new physicists. The former perpetuated their game, believing the absolute to be attainable, while they did not realize they were gazing at shadows and that they would continue to do so; the latter are fully aware of their shadowy world, yet their game has suffered no loss of self-motivation as a consequence. In fact it became much more interesting. Some contemporary scientists have become "mystics"

precisely because of the failure of physics, not because of its similarities to mysticism. Physics apparently cannot transcend the shadows, nor can we.

Another possible reason contemporary scientists have been attracted to a mystical or quasi-mystical attitude is the increasing implications, in mathematics, relativity, and quantum theory that point to the infinite. For example, the "reality" behind the shadows Jeans speaks of, that is, the universe in its totality, has occasionally been related to its counterpart: the set of all possible sets. If we begin with the empty set—{ }—and build up—$\{1\}, \{2\}, \{3\} \ldots \{n\}$—no matter haw far we progress in a lifetime, in ten, or in one thousand lifetimes, we shall never attain the highest set; surprisingly we shall remain as far removed from our goal as we were the day we began. This bears on what in set theory is called the "principle of reflection": given any possible description of the absolute, there will always be, paradoxically, a part of that absolute that corresponds to the description—which should recall Borges's use of sets in "The Library," "The Aleph," "al-Mu'tasim," and other stories.

Now, the absolute I speak of is sometimes conceived of either as One or Many. If it is said to be One, then some limiting point is presupposed; if it is said to be Many, then an endless sequence of approximations is implied. The One is the cumulation of the All, which is unknowable; the Many is a sequence to which something more can always be added, so the end will never be reached. If time were to stop, and if we were immortal, perhaps we could know the cumulative One and the sequential Many. But surely we would suffer the fate of Borges's Immortal, for whom there is nothing new under the sun: unadulterated tedium.

The One is comparable to Borges's Aleph tropologue, the Many to his Zahir tropologue. The first entails the actual infinite, the second the potential infinite: the Platonic realist orb versus Aristotelian nominalism. The two poles of this Aleph-Zahir complementarity also parallel, in another manner of speaking, the logician's and mathematician's conception of truth and provability. Gödel's incompleteness theorem establishes a line of demarcation between the two. Using correct axioms from within the system, those statements that are provable will be true (consistent), but it is impossible to know which of the true statements are provable and which are not without introducing an axiom from outside the system, so if the system is consistent, it is incomplete, and if complete, inconsistent.

Chapter Five Gödel's proof is also relevant to the distinction between semantics and syntax. Our view is semantic if the statements in a language are regarded as describing a fixed universe; it is syntactic if the symbols of the language are regarded as the elements in a game to be manipulated according to a definite set of rules. "Is statement S true or not?" is a semantic question; "Is statement S provable?" is the problem of syntax. Those dwelling on the first type of problem are chiefly Platonists; those attending to the second are usually formalists or intuitionists.

Truth is like the Aleph, the absolute in one fell swoop condensed to a point (or a cipher for Tzinacán). Provability is the Zahir, the sequence, which can never be fulfilled. The Zahir is sharply distinguished from the ineffable Aleph, for there is really nothing ineffable about it in the empirical sense. It is an ordinary coin. Its ineffability resides in its infinitely extensive linear concatenation: one damn thing after another, so to speak—and this, precisely, constitutes Funes's tedium. It is like the decimal expansion of *pi,* which, for all we know, is random, in contrast to the instantaneous but impossible grasp of infinity or zero. The Aleph is Alpha, total unity, the highest dimension. The Zahir is Omega, the common ground underlying everyday existence. The narrators of both stories (which is the same narrator) exist between the one and the other, like Pascal between the infinite and the infinitesimal.[7]

Actually, a naked singularity combines the One and the Many: in an instant and at a point. It is not, in this sense, patterned merely by the Aleph; it is the Zahir as well, but significantly, since it is the Zahir in its plenitude collapsed to a timeless moment, it must entail the Zahir's completion. Hence for mere mortals the Zahir is, in its totality, as ineffable as the Aleph, or as a singularity. The Zahir is the sum of everyday experience—tiles in the backyard of Soler Street, tigers, ferns on a greenhouse floor, ants, etc.—coupled with the totalizing Alephic grasp—the circulation of one's own blood, endless mirror images, the delicate bone structure of a hand, etc.—from a higher dimension and from every possible perspective. The Absolute, then, consists of the union of these two Borgesian images.

Common ground between Borges and contemporary physicists' disposition toward mysticism is found in this Aleph-Zahir conjunction. In fact, the Aleph and Zahir imply two complementary paths to mystical insight. Rudolf Otto (1932) describes two ways in mysticism of arriving at the ultimate: inward, toward nothing (zero), and outward, toward everything (infinity). The

first involves a cessation of thinking altogether; it strives toward the absolute through the void, which underlies everything. By way of cancellation—i.e., the saying, "Neti, neti" (or "Not that, not that")—the mind is emptied of all thoughts, and thoughts about thoughts, and thoughts about thoughts about thoughts, until, hopefully, enlightenment arrives. The path toward unity entails an effort to encompass more and more into the field of consciousness, to attain an oceanic oneness with everything.[8]

In the final analysis, plunging into consciousness toward the void and spreading outward to include the All have the same goal: nothing is tantamount to Everything. Rucker (1983, 226–27) offers a geometrical metaphor for these two mystical activities. Consciousness is a circle with a radius of 1. The inward turn proceeds toward the center, a point, nothingness (i.e., $\frac{1}{2}$, $\frac{1}{3}$, $\frac{1}{4}$, $\frac{1}{5}$, . . . $\frac{1}{n}$). The outward grasp enlarges the circle in the opposite direction (i.e., 2, 3, 4, 5, . . . n). The end product of the two goals is identical, explains Rucker, for we can regard the infinitesimal and the infinite as being the same place by first shrinking the infinite circumference of the circle to the infinitesimal center and then bending it back into the shape of an infinite doughnut (i.e., torus) with an infinitesimally minute hole. (Recall that the torus is a model-metaphor of the universe, and that a singularity is the shrinking of the totality into the infinitesimal.)

This meditation on the Aleph-Zahir pair ushers in another related question: the plurality and the oneness of the self, or of consciousness, which also bears on the mystical view. Schrödinger, perhaps the most mystical of the twentieth-century physicists, provides us with a foothold. Evoking thoughts dear also to Borges—those of Schopenhauer and Indian philosophy—Schrödinger writes:

> From the early great Upanishads the recognition of ATHMAN = BRAHMAN (the personal self equals the omnipresent, all-comprehending eternal self) was in Indian thought considered . . . to represent the quintaessence of deepest insight into the happenings of the world. . . . To Western ideology the thought has remained a stranger, in spite of Schopenhauer and others who stood for it. (1967, 93)

Schrödinger concludes, with what he calls the "arithmetical paradox" (cited in the second section of Chapter Two), that consciousness is never experienced in the plural, only in the singular. Schrödinger then asks how plurality, generally considered to be

Chapter Five an intuitive given for Westerners, could have arisen. He speculates that each consciousness discovers itself immediately bound to and dependent upon a limited region of space: the body, which is separated from all other bodies in the world. The body undergoes vast changes from birth to puberty, aging and death, and with it the mind changes. Hence the plurality of bodies and consciousness appears to be the most likely hypothesis, which is almost unanimously accepted by Westerners. The intimation of plurality has led to the body/soul split, and then to enigmatic questions: whether or not souls are immortal, whether animals and women have souls, and so on. These consequences must force suspicion on us regarding the plurality hypothesis. The only alternative, Schrödinger posits, is

> simply to keep to the immediate experience that consciousness is a singular of which the plural is unknown; that there *is* only one thing and that what seems to be a plurality is merely a series of different aspects of this one thing, produced by a deception (the Indian MAJA); the same illusion is produced by a gallery of mirrors, and in the same way Ganrisanker and Mt. Everest turned out to be the same peak seen from different valleys. (1967, 95)[9]

The coincidence of Schrödinger's thought and that of Borges regarding the Many and the One, or the Aleph and the Zahir, needs, I trust at this point, no additional elaboration. Borges's fictions, commensurate with the speculations of many contemporary physicists, tackle some of the ultimate metaphysical and logical problems head-on. It is perhaps unfortunate that during the first half of this century logical positivism generally predominated; as a result many of these problems were dismissed as nonsensical in general and fuzzy-minded at best. These are precisely the problems, however, that will in all probability endure well beyond the present century.

Chapter Six What Is Real?

This chapter begins with a brief outline of the development of quantum theory: its undoing of the subject/object split, the possible role of consciousness in quantum theoretical formulations, and quantum theory's "idealist" strain as compared to the "idealism" of Tlön. The element of chance, or quantum theory's "dice-playing God," as Einstein pejoratively put it, is contextualized in Borges's "The Lottery of Babylon." Two recent conceptions of the quantum universe, in addition to the widely accepted Copenhagen interpretation, are compared with "The Garden of Forking Paths," "Pierre Menard, Author of the Quixote," and other works. Finally, and rather surprisingly, symmetries and asymmetries in certain Borgesian texts are suggestive of the "broken symmetries" recently discovered at the subnuclear level, which apparently disrupt the cherished belief in a totally harmonious universe.

Chapter Six

The world is very complicated and it is clearly impossible for the human mind to understand it completely. Man has therefore devised an artifice which permits the complicated nature of the world to be blamed on something which is called accidental and thus permits him to abstract a domain in which simple laws can be found.
—Eugene Wigner

The Most Probable World

1 Despite the fact that there are almost as many interpretations of quantum theory as there are major quantum theorists, the illusion persists in many quarters that they all speak the same language (Colodny 1972). Though there is no common language, I shall, nonetheless, attempt to present a general picture of this slowly emerging cosmology.[1]

At the beginning of the present century, the physical universe was still regarded as a huge clockwork mechanism, its working parts legislated in every detail by the inexpugnable logic of cause and effect embodied in Newtonian principles. It was believed that the most fundamental particles of the universe, atoms, behaved like objects in our perceived world, billiard balls being the paradigm example. And an individual atom was thought to look something like our solar system, with a nucleus of protons and neutrons surrounded by gyrating electrons. It was soon discovered that electrons do not behave exclusively like billiard balls bouncing off one another; they also manifest lightlike behavior, as though they consisted of waves. The fact that particles display wave properties gave evidence that they represented some kind of statistical effect, a discovery which soon led to the unending of causality and predictability. No more could be known but that a given electron had a certain probability of being at a certain place at a certain instant. This probability construal of the quantum world, now generally known as the Copenhagen interpretation, has gained widespread, though not unanimous, acceptance.[2]

To get a better feel for this world of probability, an example is in order. Some heavier atoms have unstable nuclei and tend to disintegrate, emitting radioactive particles. It is possible to calculate the probability that a given radioactive nucleus will either emit or not emit a particle after a certain period of time.

Suppose a nucleus and a particular duration of time are chosen such that the probability is fifty-fifty that after the duration there will have been an emission. We have at the end of this duration two possible states: in one the nucleus remains intact, in the other it has decayed slightly, and both of these worlds were at the outset equally likely. This seems to be a mere toss of a coin. Either heads or tails will show, but while the coin is flip-flopping in the air it is impossible to know which. However, our everyday physical world in principle is still predictable by mechanical laws. In the first place, if we know the temperature, air conditions, the force of the coin toss, the height of its arc, etc., the resting position of the coin is in theory predictable. Not so with the subatomic world, for there is no way of knowing whether or not the radioactive nucleus will have emitted a particle after the given temporal increment. In the second place, the coin, while flip-flopping, entails two alternative worlds, but in the quantum realm the two worlds are no more than a potential. We cannot speak of the existence of either until one of the two has been actualized, i.e., until it has been, so to speak, "observed," interacted upon by another "entity." Before the "observation," there is no more than an overlapping combination of "superposed" waves. After the "observation," a wave has "collapsed" into a "particle."

Schrödinger, who along with Einstein and others sought to overthrow the Copenhagen interpretation, created a paradoxical thought experiment called "Schrödinger's cat."[3] Suppose we put a cat in an enclosed box containing a mechanism triggered by the possible radioactive decay of a nucleus that is capable of breaking a flask of cyanide gas. If the nucleus decays, the cat dies; if not, it remains alive. There is at the outset, in a rather metaphorical way of speaking, an overlapping "superposition" of two possible worlds: decayed nucleus and dead cat or intact nucleus and live cat. The question is, before we open the box at the designated time, is there a potential state, two "superposed" waves, entailing two nonactualized cats: live-dead? If so, then neither is "real" until we lift the lid of the box and take a peek (interact with) its "contents." In such case, and contrary to our better judgment, we are forced to conclude that we the spectators bring one of the two worlds into existence. As Bohr (1958, 81) tells us, we are both spectators and actors in the great drama of existence—and in a certain manner of speaking, idealism wins another round over nominalism. The problem is that our

"better judgment," or "common sense," cannot be trusted where quantum phenomena are concerned. It is one thing to speak of a subatomic particle being neither here nor there but a "superposed" wave state. It is another thing, and well-nigh impossible, to imagine a schizophrenic cat that is not really "real," yet is both dead and alive. Quantum theory contradicts the most fundamental givens of our everyday existence.

The bizarre problem created by Schrödinger's thought experiment has of yet not been definitely resolved (though some merely discard it with the rejoinder that the microcosmic domain simply cannot be exemplified with macrocosmic phenomena; that is, subatomic interactions are not analogous to what we construe to be our subject/object existence). From Schrödinger's view of things, Samuel Johnson, who kicked a stone to disprove Berkeley, or Diogenes, who stood up and walked across the classroom to disprove Eleatic philosophy, confirmed nothing except that stones and the classroom floors exist only upon such interaction. Strangely enough, to the question, "What is the world 'out there' " corresponds the answer, "There is *no* world 'out there,' " a physical world as we ordinarily believe we know it, that is.

To offer a more concrete image, consider once again the Necker cube (figure 8).[4] In its two-dimensional state the "face" is neither up nor down; in fact, there is no "face," but merely a conglomerate of straight lines. The "cube," as an imagined three-dimensional construct is a potential, or metaphorically speaking, a "superposition" of two possible future states. If at any given instant you were to glance at the figure, there is no way to predict which of the possible states you would actualize. The best we can do is state that there is something in the order of a fifty-fifty probability of your cube's being either face-up or face-down, but neither cube will "exist" until you look at it.[5]

Back to Borges, who, to reiterate, asks us to admit what all idealists admit, "the hallucinatory nature of the world," and that we do what no idealist has done, "search for unrealities that confirm that nature." Such pursuit of unrealities is precisely that of the inhabitants of Tlön, whose world is merely a succession of mental processions. Jeans remarked that the universe is now being regarded as a Great Thought rather than a Great Machine. The Tlönians consider thought to be synonymous with "reality"; understandably psychology is their major discipline. We read that there is no science, as we know classical science, in Tlön. But actually there are countless sciences: the mind, each

mind, creates its own laws of nature, its own taxonomies, its own world. In order to facilitate at least some comprehension of the inconceivable thesis of materialism, a heresiarch of the eleventh century devised the scandalous sophism of nine copper coins:

> *On Tuesday, X crosses a deserted road and loses nine copper coins. On Thursday, Y finds in the road four coins, somewhat rusted by Wednesday's rain. On Friday, Z discovers three coins in the road. On Friday morning, X finds two coins in the corridor of his house.* The heresiarch would deduce from this story the reality—i.e., the continuity—of the nine coins which were recovered. *It is absurd* (he affirmed) *to imagine that four of the coins have not existed between Tuesday and Thursday, three between Tuesday and Friday afternoon, two between Tuesday and Friday morning. It is logical to think that they have existed—at least in some secret way, hidden from the comprehension of men—at every moment of those three periods.* (L, 11)

Defenders of common sense maintained that this paradoxical anecdote was a verbal fallacy lacking in rigorous thought. The verbs "find" and "lose," they claimed, were used illegitimately. The coins supposedly having existed from the instant they were lost to the moment of their rediscovery would imply their continuous existence—the view of classical Western science—which was intuitively impossible for the Tlönians. They believed the coins ceased to exist once they were lost, i.e., unperceived, and popped into existence upon their being found. Idealism ruled—and the furniture of Tlön was presumably discontinuous: *being was* only upon *being perceived*. Or, in the quantum theoretical sense, a set of "superposed" waves is actualized into one of a number of probable events upon interaction.

Moreover, we have here, Borges tells us, the Tlönian equivalent of Zeno's arrow paradox. Is the arrow stationary during each increment of time and space, and is there no movement from one increment to another? A positive response would be out of the question from the viewpoint of the Westerner's experience, but it is logically valid within Zeno's framework. In contrast, for the idealistic Tlönian metaphysicists, intuition dictates that the coins cease to exist when they suffer the absence of their subject's gaze. In other words, the Tlönian's world is "a heterogeneous series of independent acts" (L, 8), much like the series of synchronic states of Zeno's arrow.

Chapter Six   The Westerner ordinarily senses that time is a flow, a "concourse of objects in space" (*L*, 8). Objects in the world, he believes, are an ensemble enjoying continuous beingness and self-identity through time. And through time, he is able to connect a given fact with the temporal stream of previous facts. Such linking is for the Tlönians unthinkable; linkage of one fact to another occurs in a later mental state "which cannot affect or illuminate the previous state. Every mental state is irreducible: the mere fact of naming it—i.e., of classifying it—implies a falsification" (*L*, 9–10). That is, in a manner of speaking, to name a thing is to say that, in the *now* of the thing's naming, it is *not* what it *was*. Then to connect its *nowness* with its *wasness*, another temporal increment is required, and then another, to connect the previous three, and so on. Bradley once again makes his presence felt here.

In Tlön, consequently, there are only particulars, and understandably, an undefined plurality of sciences proliferates. Each science implies its own falsification, which demands another conjecture, another science, and another almost-instantaneous refutation. In short, the Westerner's world is a continuous succession, the Tlönian's a discontinuous series. Zeno's construct is an absurdity to the experience of the former, it is simply the way the world is for the latter.

The contrast between this postulated Westerner's intuition and that of the Tlönians is, quite strikingly, parallel to the contrast between our commonsensical assumptions, on the one hand, and quantum "events," on the other. Quantum "events" are not "real" until a wave packet has been "collapsed" into a "particle" as a result of interaction, and when such interaction ceases, the "particle" reverts to a set of superposed waves, much like the coins in the discontinuous world of the Tlönians. (However, there is a fundamental difference: the interconnectedness of the quantum universe, discussed in the fourth section of this chapter, in contrast to the autonomous particulars of Tlön.)

In another sense, the Tlönians' predicament is, however baffling to our mind-set, that of Schrödinger's cat and the inconceivable behavior of the quantum world. "Idealism" of the Schrödinger cat variety is the only plausible world according to Tlönian metaphysics; hence the heresiarch's vulgar "materialism" is outlandish. All sciences, all metaphysical doctrines, compose a dialectical game of "as if," a set of "incredible systems of pleasing design or sensational type" (*L*, 10). Metaphysics is "a branch of fantastic literature," for a system is nothing more than

"the subordination of all aspects of the universe to any one such aspect. Even the phrase 'all aspects' is rejectible, for it supposes the impossible addition of the present and all past moments" (*L,* 10). It would appear, ironically, that the proliferation of quantum interpretations during the past fifty years is comparable to the Tlönians' metaphysical quest: the goal is not absolute, unadulterated Truth, but the indeterminately variable astounding.

Examples of such a quest abound. One of the most intriguing episodes involves the charismatic figure, Niels Bohr. In 1958, the physicist Wolfgang Pauli was presenting some new ideas he and Heisenberg had formulated to a group of distinguished scientists, including Bohr, in New York. A debate ensued, and as it continued, some younger scientists initiated a scathing criticism of Pauli. After some time, Bohr rose to address himself to Pauli, and as usual, the audience fell silent in respectful anticipation: "We are all agreed that your theory is crazy," he began. "The question which divides us is whether it is crazy enough to have a chance of being correct." Freeman Dyson (1958, 79–80), commenting on this occasion, remarks that the objection "that they are not crazy enough" has applied to all attempts at a new theory of elementary particles, and especially to crackpot theories, few of which get published in the most serious journals, not because they are unintelligible but precisely because they are so readily intelligible. Those that are most bizarre generally find their way to a journal. In like manner, Borges is not a radical thinker because he believes that "unreality" is the only "reality." He is radical because he incessantly demonstrates how unintelligible the very notion of an intelligible "reality" is (Ferrer 1971, 59).

In light of quantum theory, it has been evident for some time that the notion of a subject/object split is no longer tenable. The object, the Kantian *Ding an sich,* is in essence unknowable, even inconceivable; what can be known is limited to the mutual interaction between subject and object. This idea had its beginning in contemporary science with Ernst Mach, and it is explicit in the language critique of his contemporary, Fritz Mauthner. Language, argues Mauthner, allows for no absolute subject/object, inner/outer dichotomy. The outer world is simply unknowable. Like Mach, Mauthner believes that knowledge is no more than ordered experience, and like Kant, that experience gives facts but no concepts (Weiler 1970, 168–71). What is observed, as Heisenberg (1958a, 58) maintains, "is not nature in itself but nature exposed to our method of questioning." There

Chapter Six   can be no relatively unambiguous description *which does not include the describer.* This revolutionary transformation completely undid Copernicus; we are once more at the center of things, we cannot escape from ourselves (d'Espagnat 1983, 17–21). In the words of Planck (1932, 217):

> Science cannot solve the ultimate mystery of nature. And this is because, in the last analysis, we ourselves are part of nature and, therefore, part of the mystery that we are trying to solve. Music and art are, to an extent, also attempts to solve or at least to express the mystery. But to my mind, the more we progress with either, the more we are brought into harmony with all nature itself.[6]

This heightened self-awareness of the physicist is perhaps most dramatically portrayed by Eddington's graphic metaphor:

> We have found that where science has progressed the mind has but regained from nature that which the mind has put into nature. We have found a strange footprint on the shores of the unknown. We have devised profound theories, one after another, to account for its origin. At last, we have succeeded in reconstructing the creature that made the footprint. And lo! it is our own. (quoted in Heisenberg 1958b, 153)

Interestingly, Borges, speaking of Walt Whitman in "The Nothingness of Personality" and denying the existence of the individual "I," creates the image of an artist setting out to express himself and to express life in its totality, but, Borges reveals, the two expressions, in the final analysis, are one and the same (*I*, 91). More vivid yet, in the "Epilogue" of *Dreamtigers*, we read: "A man sets himself the task of portraying the world. Through the years, he peoples a space with images of provinces, kingdoms, mountains, bays, ships, islands, fishes, rooms, instruments, stars, horses, and people. Shortly before his death, he discovers that that patient labyrinth of lines traces the image of his face" (*DT*, 93).

The mind travels along mysterious unknown pathways, constantly repeating itself, each repetition being a difference, each difference disclosing a momentary glimpse of successive unknown bits and pieces of that receding horizon; each bit and piece, when gathered in with a recent harvest of the known, is a gain but necessarily accompanied by a loss, due to our finitude. Quantum theory acknowledges this finitude in its announce-

ment that the world "out there" is relegated to the status of little more than hallucination. Our very "reality" appears to have crumbled, and it has been replaced by an abstract formulation, the ramifications of which are so strange that its consequences have not yet been properly accounted for—indeed, if it is possible to account for them at all. Above all, quantum theory remains plagued by "*an essential element of ambiguity . . . involved in ascribing conventional physical attributes to atomic objects*" (Bohr, quoted in Colodny 1972, 133).[7] In fact, the quantum world is shot through with paradox. This is especially evident in the particle/wave dichotomy. It is impossible to say whether the electron is both a particle and a wave, either a particle or wave, or neither a particle nor a wave. Natural language, as will be discussed in Chapter Seven, is simply incapable of expressing quantum events. If we ask, for instance, whether the position of the electron remains the same, the answer must be negative. To the question whether its position changes with time, we must again answer no. Furthermore, two electrons can be referred to as either the same or different. If we say they are the same, we are simultaneously right and wrong: if we say they are different, we are also both right and wrong (Oppenheimer 1954, 42–43). Words seem to have lost their accustomed meaning here. This situation is strangely reminiscent of the *via negativa* of medieval mystics and of enlightenment through a Zen Koan—or perhaps of reading Borges, Beckett, or Mallarmé, watching a play by Pirandello, contemplating a work by Dalí, Picasso, or Escher—or, reading a piece by Derrida, and so on.

Quantum ambiguity culminates in the exasperating principle of uncertainty, which has enjoyed widespread acceptance since the beginning of the 1930s, when Eddington remarked that there was now no division of opinion as to the decease of determinism; if such a breach existed among scientists, it was between the "mourners and the jubilants" (quoted in Stebbing 1958, 185). Briefly, uncertainty arises from the fact that we are once again inextricably confronted by Zeno. The picture of space-time is essentially static. An electron that has an exact location in space-time, by this very fact, is deprived of the possibility of motion. On the other hand, if it is considered to be in motion, then it cannot be attached to any point in space and time; it can only be conceived as somehow impossibly "between" the static set of quantum increments making up its trajectory. Here, without doubt, "lies the most profound sense of the

mysterious concept of the quantum of action" (de Broglie 1955, 121–122).

Moreover, quantum theory, like Zeno's paradox, entails infinity, but of a different sort: Hilbert space of infinite dimensions. For example, one dimension can be taken as a single line; in order to have two dimensions, that line must be extended laterally to form a sheet; three dimensions require that the sheet be moved laterally to produce a cube. With the three-dimensional cube it is possible to reorient the lines in many different ways and still cover the same space. In fact, just as a point in space can theoretically have an infinity of lines passing through it, so the cube can be reoriented in an infinity of ways in an infinity of dimensions. Now, an isolated electron is describable by quantum theory in three-dimensional space, but for a two-particle system, six dimensions are necessary, and for 10^{24} particles, 3×10^{24} dimensions. There is theoretically no upper limit. For a given "particle," then, all knowledge concerning it can be described by its wave function, which consists of a (countable) infinity of possibilities, or numbers. Knowing all these possibilities, one can predict with what probability the particle will behave in such-and-such a way at a given instant. The problem is that there is virtually an infinity of possibilities for each state of affairs, so one cannot know with certainty.

Quantum uncertainty indirectly bears on the problem, once again, of whether Borges's Library, i.e., the universe, is finite or infinite. If the alphabet and languages of the Library's books allow for infinite generativity (in the sense of Chomsky), then there exists the potential for an infinity of books. Supposing that the total possible combinations of symbols in the books is infinite, and the Library's inhabitants cannot be in possession of any finite set of rules for generating the books (as there presumably is in Chomskyan generative-transformational grammar), then as far as they are concerned, absolute randomness must surely prevail, and if so, then the Library suffers from absolute uncertainty. The Library's denizens remain victimized by a collection of books; the randomness is totally beyond their control.

If a sort of "quantum anarchy" rules in the Library, in Tlön the opposite is the case, as evidenced by the parable of the nine copper coins, which is, in a strangely paradoxical manner, related to the concept of the observer-participant. Berkeley has his heyday in "Tlön" (in fact, the narrator tells us that Berkeley was once

a member of the benevolent secret society that invented Tlön). Since for the Tlönians to think something suffices to bring it into existence, "textuality" can become "reality" and vice versa. The literature in the northern hemisphere of Tlön abounds in Meinongian mental objects, which are "convoked and dissolved in a moment, according to poetic needs." In fact, literature obviously enjoys the same status as "reality," however ephemeral it might be, for the fact that no one believes that nouns are real "paradoxically causes their number to be unending" (*L,* 9). Nouns are no more than poetic objects, which can be conjured up at will. The director of a state prison in Tlön once told his inmates that ancient tombs existed in a dry riverbed, and he promised freedom to anyone finding them. After a few abortive attempts, "a gold mask, an archaic sword, two or three clay urns and the moldy and mutilated torso of a king" (*L,* 14) were unearthed. A doorway survived so long as a beggar continued to visit it; when he died it disappeared. We are told that even "some birds, a horse, have saved the ruins of an amphitheater" (*L,* 14). In other words, the observer brings into existence the very world she expected-desired.

The Tlönians indomitable idealism reaches cosmological magnitudes. A certain *Anglo-American Cyclopedia* (but not all of them) has an article about Uqbar, which refers to the imaginary region, Tlön. After a volume of the *First Encyclopedia of Tlön* is found, Tlön subtly begins emerging from that text to become "reality" by engulfing our world, which includes the text, "Tlön, Uqbar, Orbis Tertius." Meanwhile, the narrator forecasts that the one hundred volumes of the *Second Encyclopedia of Tlön* will be discovered one hundred years hence, which will call a third world, *Orbis Tertius,* into existence, first as text, then as "reality."

Hayles (1984, 146) remarks that this progression "implies a sequence of worlds, each calling its successor into being in an increasingly unimaginable sequence that has no end."[8] Such sequences, by combining the known with an unattainable and unknowable end point, imply, paradoxically, "boundedness and infinite regress." And there is a further problem here. Tlön begins as a member of the class of things in Uqbar, which in turn is a member of the class of things in our world that enjoy equal status with "Tlön, Uqbar, Orbis Tertius," hence the proposition "The World *will be* Tlön," is, in a manner of speaking, equivalent to the proposition "The class of all women *is* a woman." In another way of stating this Russellian paradox, the elevation of

Tlön to the status of a general class is comparable to the barber of Seville, who advertised that he shaved those, and only those, who did not shave themselves. If he decided to shave himself, he would elevate himself to the level of the class of all those who shave themselves, and hence he could not, logically speaking, shave himself. If Tlön *becomes* the world, assuming the world to be "real"—and our propensity to assume it to be so persists, realists that we would like to be in spite of ourselves—then it cannot be a fiction. But if the world is a text rather than "real," the same quandary holds, for Tlön still began as a member of the world-text.

The problem becomes even more ticklish. On Tlön, one conjecture had it that there is only one subject, which is indivisible and at the same time all beings, and which thought (created) the universe. This subject, apparently, is tantamount to James Jeans's Great Thought. Or perhaps this venerable subject is Borges himself for, after all, he is the author of "Tlön." In this sense, Borges penned "Tlön," which contains Tlön, and which in turn becomes the world. So Borges must be the equivalent of the Great Thought. But this is logically impossible, since Borges is a member of the set of all things *in* the world. In other words, assuming the world plus the Great Thought (= Borges) to be a set, and following Russell, Borges cannot be the set of things of which he is a member.

A well-intentioned reader might offer a rebuttal: "I read the text, which includes everything you are speaking of, which is nothing but words, so I remain outside." I suppose so, if we limit our words to the nuts and bolts of everyday living. But not so if Borges has his way, and even to a degree if we take quantum theory at face value. According to Borges, Don Quixote is a reader of the *Quixote* and Hamlet a spectator of *Hamlet,* hence the general reader takes on the same status as fictional characters. "But surely this is no more than tropological play, jolly good fun, which makes for enjoyable reading, but it is hardly relevant to the hustle and bustle of the real world." Perhaps, and perhaps not. The problem is that we find ourselves ultimately returned not only to the quantum theoretical observer-observed dilemma but also to the problem of consciousness *of* oneself and *of* a world "out there." Physicists have been unable to ignore consciousness totally, though many persist in conveniently pushing it under the rug. Physics, to reiterate, does not simply describe nature. It is part of the interplay between nature and the physicist. Quantum phenomena can only be comprehensible in terms of links in a

chain of processes at the end of which is to be found a Tlönian-like observer. In this light, physics should not, or perhaps cannot, ignore the role of consciousness.

Eugene Wigner is the chief advocate for the inclusion of consciousness in the quantum picture. Drawing his inspiration from Schrödinger, he argues that the quantum world is not separable from living systems; cats, as well as ourselves, influence collapsing wave functions and specific outcomes. Consciousness, from the most rudimentary to complex levels, becomes a factor; it is firmly pegged at the center of its cosmos, by and large determining what is observed. There is, therefore, a sharp distinction between the "reality" of consciousness and the "reality" of everything "out there" (Wigner 1970, 185–99). This notion evokes the body-mind problem, but Wigner sees no dilemma here. The traditional answer to the question, "Does the mind influence the physico-chemical condition of the brain?" is, "No, the body influences the mind but the mind does not influence the body." Wigner gives two reasons for believing the contrary to be true: (1) the law of action and reaction (if the body acts on the mind, the reverse should also be the case); and (2) the fact that quantum events are altered by the observer. He concludes that, judging from the available evidence, "there exists only one concept the reality of which is not only a convenience but absolute: the conduct of my consciousness, including my sensations" (Wigner 1970, 189). Wigner then cites Schrödinger, from *Mind and Matter* (1967, 100): "Would it [the world] otherwise [without consciousness] have remained a play before empty benches, not existing for anybody, thus quite properly not existing?" (brackets Wigner's). Wigner, of course, answers in the negative, and he continues with the speculation that if we deny the absolute "reality" of objects such as a book or attribute to them a different type of "reality" from that of sensations, we are not in any way in conflict with the fact that we continue to act as if these objects were "real." This is necessarily the case, for "the usefulness of the concept of objects is so great that it would be virtually suicidal to refuse using it, in one form or another" (Wigner 1970, 191).[9] Our naive realism tends obstinately to prevail, in spite of quantum theory's suggestion that we have been Tlönians all along. Or perhaps quantum theory *is* Tlön, which is busily engaged in taking over our naive realism.

John Wheeler (1980), with a grandiose speculative leap, takes Wigner's hypothesis a giant step further, introducing the mind (observer) as coparticipant with the very creation of the

Chapter Six   universe. He suggests that the mechanism allowing the universe to come into being is unworkable unless the universe is already guaranteed to produce observers, since they are necessary for the very existence of the universe. Wheeler proposes that there must be some principle enabling the universe to spring forth of its own accord, a "self-causing" universe, similar to a "self-excited" electrical circuit. Wheeler's speculations entail a "delayed-choice" principle by means of which the observer, who observes what happens, irretrievably influences (even determines) something that has already happened. The past has no existence except in the manner in which it is observed in the present. Tangible "reality" is afforded the universe not only now, but back to the very beginning! (Recall, for example, Borges's prisoners in Tlön, who eventually discovered artifacts in their diggings because they desired/expected them to be there, and Kafka, who influenced Cervantes equally as profoundly as Cervantes influenced Kafka.)[10]

Wheeler's "tail-chasing dog" model could easily be tossed in the trashcan were it not for a number of (actual, not thought) experiments, the most noteworthy of which is that of the French physicist Alain Aspect and his team at the University of Paris-South in 1982. In brief, this experiment was a response to the so-called Einstein-Podolsky-Rosen (EPR) paradox, a thought experiment suggesting that if a pair of particles possessing a common origin and opposite "spins" fly apart in opposite directions, they remain correlated with one another, almost as if they were endowed with telepathic powers. Consequently, after the particles have been separated by a great distance, if the "spin" of one particle is altered from "left" to "right," then the "spin" of the other will be automatically and instantaneously altered from "right" to "left," an impossibility in the Einsteinian universe, which limits the travel of information to the finite velocity of light.[11]

This is the converse of a situation in which two coins are tossed repeatedly, for there is apparently no *necessary* correlation between them. Suppose, however, that somehow the coin tosses were not purely random, so that when heads showed up on one coin the other invariably showed tails. The neutral observer would soon note the correlation, which contradicts his commonsense expectations. Such contradiction is like the magician's glove trick. He keeps one and gives the other to a person in the audience. Then the magician turns his glove wrong-side

out, and to the spectator's surprise the same has automatically happened to his glove. Aspect's experiment demonstrated the impossibility of the EPR formulation that particles such as an electron pair, once together in the same atom, remain part of a single system no matter by what distance they might be separated. The upshot of Aspect's experiment, as Wheeler notes, is that the atoms, for example, of this book consist of particles that were once neighbors to other particles in the primordial glob, part of which might now be in distant galaxies.

Interestingly, Davies (1983, 111) refers to Wheeler's self-contained, self-referential, self-perpetuating "excited circuit" universe as a giant Hofstadterian "strange loop," within which microscopic "reality" is inseparable from macroscopic (the observer's) "reality." Yet the macroscopic is made up of the microscopic—the brain is itself a conglomeration of particles that, so to speak, have separated themselves from the universe in order to see *it,* which is *them.* This notion brings the fictiveness of the world to bear on the textuality of "Tlön"—and it reevokes Russell's paradox. As a fictive alternative to the world, "Tlön is necessarily set apart from the world, but at the same time, "Tlön," its author, and its readers are members of the set of all things included in the world, which is being engulfed, paradoxically, by Tlön. To paraphrase the final remarks of "New Refutation" within the context of the present disquisition:

The world is a river which sweeps Tlön along, but Tlön is the river; the world is a tiger which destroys Tlön, but Tlön is the tiger; the world is a fire which consumes Tlön, but Tlön is the fire.

Fiction, unfortunately, is what is real; the world, unfortunately, is Tlön.

The import of Russell's set-theoretical paradox to Tlön now becomes paramount: *the whole is forcibly, but paradoxically, a member of itself.* As a result of Aspect's experiment, and if Wheeler's speculations hold some validity, then a given observer's actualization of "reality," from the beginning into the future, is no less determinate than the "reality" of the Tlönians. Perhaps the universe was preordained from the "beginning," its cosmic spectator-participants merely acting out a drama that could not have been other than what it is. Or perhaps, each spectator-observer, like the fabled Tlönians, carves out her own "reality." The ultimate ramifications of quantum theory (or

Chapter Six

"Tlön") apparently raise more questions than they resolve, questions that, in and of themselves, are apparently irresolvable. This, of course, is appropriate for fictive constructs.

Significantly enough, Rucker (1984, 193), after quoting one of the more radically Berkeleyan passages from Borges's "New Refutation," goes on to remark that it is utterly astounding that such a seemingly perverse view "is embraced by modern physicists," concluding that the "kinds of questions one asks—and the order one asks them in—has a profound influence on the answers one gets, and on the world view one builds up." With this in mind, I turn to another world that, unlike Tlön, is probabilistic through and through: "The Lottery of Babylon" (L, 31–35).

The ancient covenant is in pieces; man knows at last that he is alone in the universe's unfeeling immensity, out of which he emerged only by chance.
—Jacques Monod

A Throw of the Dice

2 "Ask any molecule," writes Wheeler (1980, 352), "what it thinks of the second law of thermodynamics and it will laugh at the question." Or, ask a molecule if it is moving at random, and it will say that it is not, that it has free will (Park 1980, 60). If we observe this molecule's behavior among a collection of like molecules, it will appear to us chaotic, but if we place 6×10^{23} of them in an enclosed container, as an aggregate they become relatively predictable in terms of temperature, pressure, and volume following Boyle's law of gases. The question persists: If the behavior of one molecule is random, but when in the company of many like molecules the aggregate is predictable, then is the universe ultimately determinable or indeterminable? The latter is most likely the case, for, as Heisenberg (1966, 132) reveals, physicists now "realize more and more that our understanding of nature cannot begin with some definite cognition, that it cannot be built on such a rock-like foundation, but that all cognition is, so to speak, suspended over an infinite abyss"—a rather remarkable statement, in view of current deconstructive practices in literary criticism and philosophy.[12]

Borges is a strange combination of Apollo and Dionysus: on the one hand a determinate and reasoned game-player, on

the other a Mallarmean dice thrower—and it is precisely this second Borges who interests us here. A die thrown once affirms *chance,* thrown many times, and the throws in their composite affirm *necessity* (based on a probability factor, of course). Put one thousand identical dice in one thousand identical dice-throwing machines and throw all of them at once. Record the results. Then take one of the dice and throw it one thousand times from one of the machines. The results will be very close, perhaps even identical. The first experiment is an *ensemble,* the second is *history.* The first is analogous to the universe as succession (or the Zahir), which is indeterminate; the second is analogous to the universe as a totality (or the Aleph condensed from the Minkowski "block"), which is determinate. How do one thousand unconnected events combine to produce a predictable outcome? How does one die without memory give the same predictable pattern? Why is the ensemble average interchangeable with the time average? How can the accumulation of purely *chance* events culminate in *necessity?* Significantly, what Nietzsche calls *necessity* is not "the abolition but rather the combination of chance itself" (Deleuze 1983, 26).

Like the ultimate laws of the universe, the laws of chance remain outside our grasp. In this respect, Borges alludes to the Kabbalists, who

> thought that a work dictated by the Holy Spirit was an absolute text: a text where the collaboration of chance is calculable at zero. The portentous premise of a book that is impervious to contingency, a book that is a mechanism of infinite purposes, . . . led them to other easily ridiculed exegetic rigors. The apology of such a premise is that nothing can be contingent in the work of an infinite intelligence. (*OI,* 128; also *SN,* 130)

What does Borges mean by an infinite intelligence? He gives an example: "[T]he steps a man takes, from the day of his birth to the day of his death, trace an inconceivable figure in time. The Divine Intelligence preconceives that figure at once, as man's intellect perceives a triangle. That figure (perhaps) has its determined function in the economy of the universe" (*OI,* 128n).

This image, in light of Chapter Four, is comparable to a static, transcendent view of the "block" universe in contrast to the "time line" of an immanent, individual entity. It is also related to the distinction between an ensemble of dice throws as

opposed to a linear sequence of throws. The time-bound (historical) being, from the larger view, can appear to be the product of necessity, but as far as he is concerned, he exercises free will—as did Wheeler's molecule, with a little anthropomorphic imagination. But this being can never really know, for the larger view is inaccessible to him. That is to say, Borges's Library exists *ab aeterno*. From an immanent perspective, the Library is the equivalent of God: books are generated randomly until the canonical works show up, hence order should prevail. But the Library's inhabitants can perceive no order, since the books appear to be no more than the random juggling of letters, perhaps to infinity.

This vision of the universe as a chaotic Library when viewed from within complements the representation of life as a Cosmic Lottery in Borges's "The Lottery of Babylon" (*L*, 30–35). If from within "The Library" what appears to be worldly chaos is actually the work of a Divine Being, in "The Lottery" the game of chance determining humanity's destiny is the result of "sacred drawings," divine throws of Cosmic Dice. But if the Lottery is "the basis of reality, an intensification of chance, a periodical infusion of chaos in the cosmos," and if in reality "*the number of drawings is infinite*" (*L*, 34), then how can there be any order whatsoever, from a finite perspective? And if there is, how can it be known?

The narrator of "Pierre Menard, Author of the Quixote" (*L*, 36–44) speculates on the possibility of Menard's writing the same book as Cervantes being calculable to zero. But, he concludes, it remains nonetheless a possibility. The same can be said of the books in the Library, which might well have been typed by the proverbial army of immortal simians. What are the chances of *Quixote* being generated? Next to nil, but it yet remains a possibility, and no law of physics would be violated if, say, ape number 6×10^{23} happened to create Cervantes's masterpiece.[13] The library of books these simians might write would possess some sort of order, but, given its magnitude, the number of intelligible books would be virtually infinitesimal in comparison to the total number of books. Quite understandably, to the immanent, mortal observer, this library would appear chaotic.

The American philosopher C. S. Peirce (1960, 2:661–68) was confronted by a comparable dilemma. A pioneer in probability before it became fashionable, Peirce began with the reasonable premise that if a die is thrown, there is a probability of one in

three that in the long run a number divisible by three will turn up. What is meant by "in the long run"? The only response can be: infinity, for we cannot know if the probability is indeed one-third without exercising an endless series of trials. Since this is impossible, a physical, if not a logical, impossibility, we must remain eternally uncertain. The problem of the die turning up either a three or a six cannot be absolutely determinate on the basis of a finite series of throws, just as it is not impossible mathematically that the die will not obstinately turn up a four each time during a series of one thousand throws. We must return to this problem, but first an additional word on the Lottery.

Borges's Lottery appears to contradict the Babylonians' desire for order, for they were most fond "of logic and even of symmetry" (*L,* 32). But perhaps to ward off boredom in the face of this stifling sameness, they resorted to a game of chance. The lower classes first became obsessed with the Lottery, buying chances for a few cents in hopes of winning a relatively paltry prize. Eventually, disinterest set in, and it became necessary to increase the risks and magnify the prizes. As the Lottery became institutionalized, it was now called the "Company," and new rules were added to the game. Those who held unlucky numbers, in addition to their loss of investment, had to pay fines, and if they did not do so, they were sued or thrown in jail. Later, the losers were not fined but immediately incarcerated, and finally they were compelled to face mutilation and even death. By this time, the Lottery had become secret and obligatory for all.

Someone conjectured that if the Lottery "is an intensification of chance, a periodical infusion of chaos in the cosmos, would it not be right for chance to intervene in all stages of the drawing and not in one alone? Is it not ridiculous for chance to dictate someone's death and have the circumstances of that death—secrecy, publicity, the fixed time of an hour or a century—not subject to chance?" (*L,* 34).

The narrator admits that these speculations had brought about considerable reform, the complexities of which only a few specialists understood, but he nevertheless attempts to summarize them in a symbolic way. The Lottery, he tells us, became a Lottery-within-a-Lottery. Each loser was forced to draw nine numbers: one might penalize him, another grant him freedom, yet another give him a prize, and so on. As the number of possible drawings eventually became infinite, no decision could

ever be final; they all became ramifications of others in a dizzying labyrinth. (We need only recall Peirce's dilemma to comprehend the import of this situation.)

Some of the ignorant persisted in their belief that infinite drawings would require an infinite time, but they were unaware of the fact that "it is sufficient for time to be infinitely subdivisible, as the famous parable of the contest with the tortoise teaches" (*L,* 34). This infinity harmonizes nicely with the "sinuous numbers of Chance and with the Celestial Archetype of the Lottery, which the Platonists adore" (ibid.). Once again Borges paradoxically fuses the drawings (a series) with infinite divisibility (a continuum).

The mind generally abhors chaos and strives to find order, sometimes even when there is none: the Greeks saw mythical figures in the random patterns of the stars. Tea leaves and entrails of animals have been used to foretell the future, the gods have been consulted by casting bones, and a toss of sticks or coins is used in the *I Ching.* If randomness strikes fear in the human psyche, it also fascinates. Yet, as has been illustrated, there is no rule for determining randomness in a numerical series. But there is a method for potentially determining if a series is *not* random: simply continue the series until finally some semblance of order appears. However, when extending an apparently random series, instead of growing more satisfactory, it tends to grow less so, until at infinite length it becomes absolutely contradictory, for in an infinite series the impossible is bound to occur (Spencer-Brown 1957, 56).

What bearing does this have on the Lottery? If, as the narrator tells us, the number of drawings is infinite in the Lottery, if all possibilities branch into one another, and if no decision can be final, then it can be demonstrated that every possibility will eventually transpire. Consider a variation of what has been called the "lottery paradox" (Rescher and Brandom 1979, 45–46). If a die is thrown and the prediction made that i will *not* turn up, where i is 1, 2, 3, 4, 5, or 6, we can say with inductive certainty that there is approximately an 87 percent chance that the prediction will be true; that is, i will turn up with a 17 percent probability. Yet, if the prediction is made that all six of the numbers ($i = 1$ through 6) will *not* turn up, and if the die is tossed once, then there is a 0 percent chance of the prediction being true. A prediction for each of the six numbers considered separately with an isolated die toss enjoys a certain probability, but all six predictions conjoined with respect to one die toss

yields a zero probability. What, in essence, is the problem here? It is comparable to the distinction, as the above example of one thousand dice illustrates, between a series of single tosses, which have a history, and the simultaneity of all tosses, which is an atemporal ensemble.

To embellish this analogy further, assume we have a lottery wheel with one hundred divisions. Each time we spin it, after choosing a number, we will have a 1 percent chance of winning and a 99 percent chance of losing. This indicates that if we predict with a given spin that a particular number will *not* be the winner there is a 99 percent probability of our prediction being correct. Now, suppose we construct a lottery wheel with an infinity of divisions. If we spin it once and predict that the pointer will stop, say, on 5, our chances of winning are infinitesimally small. In order to win with a degree of certainty, we must construct an infinity of such lottery wheels. So let us do so. Now we have both an ensemble and history. Either by spinning one wheel an infinity of times or all the wheels at once, we are sure to win, no matter which number we select. This situation is comparable to Borges's Lottery with an infinity of possible drawings that do not require infinite time, if time is indeed "infinitely subdivisible." In such case, with respect to our lottery wheel, let us divide one-hundredth of a second into an infinity of increments and spin one wheel once during each time duration. This is history. Or, spinning all the wheels in simultaneity, we have the ensemble. For practical purposes, however, one-hundredth of a second can be considered as simultaneity, limited beings that we are. In this event, history is packed into the same framework as the ensemble, which is to say that if a player plays the ensemble of wheels, her number, no matter what she chooses, is certain to turn up on at least one of the infinity of wheels, and if she plays the history wheel packed into one-hundredth of a second, she will still win during what she conceives to be, for practical purposes, an infinitesimal increment of time. Absolute uncertainty becomes absolute certainty, unadulterated chance becomes necessity.

The Babylonians, caught within this type of dilemma, attempted to falsify the inevitability of infinitely contingent happenings that ultimately become necessity. Their historians, who were "the most penetrating on the globe," invented a generally reliable, though devious, method for correcting chance. But ultimately, the attempt to subvert the infinite disorder, which is also tantamount to an infinite order, fell into the hands of the

Chapter Six　　Company, for in an apparently chaotic world where the "drunkard who improvises an absurd order, the dreamer who awakens suddenly and strangles the women who sleeps at his side, do they not execute, perhaps, a secret decision of the Company?" (*L,* 35). Such must be so, since if all possibilities are certain to occur, either in the "now" (of the ensemble) or at some "point" (in history) in this infinite game, then everything must be a decision of the Company. There is a number of hypotheses concerning the resultant state of affairs. One, that the Company ceased to exist centuries ago, and now the "sacred disorder" of the lives of the Lottery's players is purely hereditary, another that the Company is eternal and will endure until the last god "annihilates the world," and still another, that the Company *"has never existed and will never exist."* The most interesting hypothesis posits that "it is indifferent to affirm or deny the reality of the shadowy corporation" (*L,* 35). Indeed, either an affirmation or a denial of the Company must be met with indifference, since nothing can be either correct or incorrect in this game of chance. Nor is it possible for anyone either to win or to lose, nor does everybody both win and lose.

In other words, we have returned to the problem of the One and the Many. From the omniscient gaze of the Company, the One, the Lottery is ensemble: all things must occur. The Babylonians, victims of immanence, are helpless and hopeless gamblers inside history, the Many. Yet since the Company introduced infinity into the game of chance, everything must occur in history as well, hence, given Babylonian collectivity, there are no losses to be suffered. The problem is that they do not, and cannot, know this. Of course, the Company must also be in total control of history as well, as was our gambler using an infinity of lottery wheel spins packed into one-hundredth of a second.

In this light, consider Escher's words, which create a complementary image of history and ensemble:

> Anyone who plunges into infinity, in both time and space, further and further without stopping, needs fixed points, mileposts, for otherwise his movement is indistinguishable from standing still. . . . Anyone who wishes to create a universe on a two-dimensional surface (he deludes himself, because our three-dimensional world does not permit a reality of two nor of four dimensions) notices that time passes while he is working on his creation. But when he has finished and looks at what he has done, he sees something that is static and timeless; in his picture

no clock ticks and there is only a flat, unmoving surface. (Escher 1971, 15)

What Is Real?

Escher's woodcut is successively and indeterminately elaborated. This is history. When completed, as a totality, it is a timeless, determinate entity. This is ensemble. The artist, of course, forges order from chaos, necessity from chance. But in its composite, the infinite randomness to which I have referred regarding the Lottery is tantamount to Escher's static, timeless order. Of course, the polar opposite of order is traditionally thought to be disorder or randomness. An example is found in quantum randomness, which appears to be absolute. Radioactive material, as witnessed by Schrödinger's cat, decays in a totally random fashion, which at the same time calls up the universe—a timeless order—to actualize its event. Occasionally a flaw can be detected in a gambling machine, because it is invented and fabricated by mere mortals, hence the machine's randomness is not absolute. In contrast, physicists have found no flaw in the randomness of the quantum world. If dice-playing God there be, he/she is most probably an honest gambler. In this light, the "Company" is not the construct of a mischievous demon who buffets the lottery players about at will. It is the product of infinite chance, which in its composite paradoxically becomes the equivalent of necessity.

Everything divides up into itself, I suppose.
—*Samuel Beckett*

3 In 1953, with Wheeler's encouragement, Hugh Everett III proposed for his doctoral dissertation what remains perhaps the strangest and most novel interpretation of quantum mechanics, now known as the "many-worlds interpretation."

Multiple "Realities"

An analogue image of this interpretation, I hope to illustrate, is found in a temporal labyrinth, the infinite book by Ts'ui Pên described in "The Garden of Forking Paths" (*L*, 19–29), which I shall introduce before turning to Everett's quantum formulation. From one perspective, Borges's tale is a spy story. It is also much more. Dr. Yu Tsun, an agent for the Germans during World War I, is in England expecting execution, for he knows the British intelligence is in hot pursuit. His objective is somehow to

Chapter Six   inform Berlin of the name of the town the English are preparing to attack, but the normal lines of communication have been broken. Realizing he cannot escape, he devises a plan, which does not become evident to the reader until the end of the story. He picks out a name in the telephone directory identical with the name of the town to be invaded, knowing that if he can kill this person, he in turn will soon be executed, and the two names will appear in the newspaper the following day, providing his allies with the necessary clue. As he leaves in the train for the residence whose owner bears this name, he catches a glimpse of Madden, the agent trailing him. Time is pressing. After temporarily evading Madden, Yu arrives at the home of Stephen Albert, who, coincidentally, was a former missionary in China and a dedicated sinologist. Yu is invited in. The two engage in an amicable discussion, which eventually leads to Yu's ancestor, Ts'ui Pên, an ancient Chinese astrologer and writer who had composed an extraordinary book, *The Garden of Forking Paths,* which Yu and Albert discuss at length before Yu kills him. Madden, now approaching the house, quickly arrests Yu, who is then condemned to die at the gallows, his mission having been completed.

Ts'ui Pên had once set for himself an inconceivable task: to construct an infinitely complex labyrinth and to write an interminable novel. At his death it was considered by all that he had failed, for a labyrinth was nowhere evident and his unfinished novel was virtually nonsensical (characters would die, only to appear later, and other such absurdities). Albert revealed that he had discovered the key to this enigmatic novel: just as the only prohibitive word in a riddle whose answer is chess is the very word chess, so also for this novel, which is a gigantic riddle wherein the word time does not appear. The novel *is* the labyrinth, and the labyrinth is temporal rather than spatial. Time in the novel does not unfold linearly, but bifurcates into multiple directions, revealing the ramifications of all possible events in their myriad totality: those events which occurred as well as those that did not, all of them coexisting as alternative realities. In other words, Albert tells Yu:

> *The Garden of Forking Paths* is an incomplete, but not false, image of the universe as Ts'ui Pên conceived it. In contrast to Newton and Schopenhauer, your ancestor did not believe in a uniform, absolute time. He believed in an infinite series of times, in a growing, dizzying net of divergent, convergent and parallel times. This network of times which approached one another, forked, broke *off,* or were unaware of one another for centuries,

embraces *all* possibilities of time. We do not exist in the majority of these times; in some you exist, and not I; in others I, and not you; in others, both of us. In the present one, which a favorable fate has granted me, you have arrived at my house; in another, while crossing the garden, you found me dead; in still another, I utter these same words, but I am a mistake, a ghost. (*L*, 28)

Princeton University Press published Everett's doctoral dissertation along with a collection of papers on it entitled *The Many-Worlds Interpretation of Quantum Mechanics* (Dewitt & Graham, 1973). Most appropriately, the above Borgesian quote from "The Garden" appears in the volume as an epigraph.[14] In the words of Bryce DeWitt (1973, 161), who has worked with Everett's interpretation of quantum mechanics since the late 1960s, the universe of the many-worlds interpretation, like the above passage depicting Ts'ui Pên's labyrinth, "is constantly splitting into a stupendous number of branches, all resulting from measurement like interactions between its myriads of components. Moreover, every quantum transition taking place on every star, in every galaxy, in every remote corner of the universe, is splitting our local world on earth into myriads of copies of itself."

Briefly, the many-worlds universe is a self-contained cosmology described as an unimaginably complex superposed wave function. There is no outside observer (God) who collapses this universal wave function representing all potential alternative realities. The collapsing comes from within. Reconsider, for instance, Schrödinger's cat. While the box is closed, the cat is represented by the superposition of two waves: live-dead. Neither of the two is actualized until the box is opened and observation (i.e., interaction) occurs. Suppose the cat is seen to be alive. Well and good. But what about the other possibility, dead cat, that previously existed? According to Everett's formulation, though this alternative world was not actualized by this particular observer, there is no cause for considering it to be any less "real." The dead cat is every bit as "real" as the live cat. It branches off into its own parallel universe. Reducing this macroscopic cat model to the microscopic quantum level, the many-worlds interpretation implies that at each instant each particle is bifurcating into parallel universes.

Everett's hypothesis reads like science fiction, but it goes much deeper, and it is based on elegant mathematical equations, consistently and logically working out some of the puzzles of the quantum picture that remain beyond the limitations of

other interpretations. Wheeler (1957, 464), enthusiastically supporting Everett, declared that it is difficult "to make clear how decisively the 'relative state' formulation [of the many-worlds interpretation] drops classical concepts. . . . [T]his step can be matched but a few times in history." And DeWitt (1973, 161) recalls the shock he received on first reading the concept: "The idea of 10^{100+} slightly imperfect copies of oneself all constantly splitting into further copies, which ultimately become unrecognizable, is not easy to reconcile with common sense. Here is schizophrenia with a vengeance."[15]

Though the imagination tends to run wild when one reads about this interpretation, certain caution should be exercised. The hypothesis specifies that the universe branches into all *possible* ramifications, but that *pure imaginables* (i.e., Borges's "imaginary beings," fantastic stories, etc.) are not among them. Moreover, though Everett's model can justifiably be called, following Deleuze and Guattari (1983), "unbounded schizogenesis," and mind-boggling though the number of branches may become in these multiple universes, they remain, nonetheless, finite in number, though they might appear infinite to a finite being (this is a good lesson for those who speculate on the magnitude of Borges's Library). In addition, Everett warns that any chance of our taking lateral moves which transport us into alternative universes that could have been actualized in our own universe but were not is out of the question. Parallel worlds, once disconnected, are physically isolated from one another (Davies 1983, 116–18).

Walter Mignolo (1977, 309), on discussing "The Garden," refers to DeWitt's article and then goes on to construct a "branching tree model" of Ts'ui Pên's book using "temporal logic" in order to demonstrate actualized linear time in contrast to the unactualized branching possibilities. Though Mignolo does not relate Borges's work specifically to the many-worlds interpretation, he rightly concludes that the branching model of temporal logic and the branching structures of Borges' stories, upset the Newtonian-Kantian concept of temporal dimensions. . . . The difference between the two lies in that the Kantian model permits a *possible configuration of events* (a history), while the branching model, a *configuration of possible events* (various histories).

The first is a linear-temporal model (i.e., history, according to Borges's Lottery); in the second, everything possible happens at once, reminiscent of Minkowski's "block," or the ensemble.[16] A problem persists, however. If the universe by definition is self-contained, with no outside observer, and if

quantum waves cannot be collapsed (i.e., the furniture of the world cannot come into existence) without an observer, then how did it all begin? Wheeler's idea of consciousness functioning by means of reverse causality to the primordial big bang is one answer, but it entails a circular argument as perplexing as the dilemma it is supposed to explain away. Neither is Wigner's "solipsist" argument of the sole observer in the universe, "myself," creating reality from the range of quantum possibilities a comforting metaphysics. On the other hand, the many-worlds model of quasi-infinite universes existing in some sort of "superspace" and "supertime" strikes one as utterly unthinkable, but the most beautiful of all. The universe is conceived as a quasi-infinity of Chinese boxes: an observer opens the box and observes Schrödinger's cat to be alive, another observer observes the first observer observing the cat, a third observer observes the second observer, and so on. The entire universe lifts itself up by its own bootstraps. Perhaps reality is not only stranger than we think, it is stranger than we *can* think.

When considering the universe to be Ts'ui Pên's novel — which is the implicit intention of the narrator in Borges's short story—we can interpret it as the atemporal "block" that contains, in simultaneity, our universe plus all its innumerable alternatives coalesced into a moment. Yu, when striving to grasp the meaning of his ancestor's novel-labyrinth, thinks of it as a "labyrinth of labyrinths, of one sinuous spreading labyrinth that would encompass past and future" (*L,* 23). Absorbed in his "illusory images," he felt himself to be "for an unknown period of time, an abstract perceiver of the world" (ibid.), an experience comparable to Borges's "feeling in death" as well as that of other characters such as Tzinacán and the narrator of "The Aleph" when they witness the marvelous. Yu, at the termination of his conversation with Albert and commensurate with the concept of the many-worlds interpretation, senses a "swarming" jungle of quasi-identical twins existing within their niches of time and space, in their "parallel" worlds: "It seemed to me that the humid garden that surrounded the house was infinitely saturated with invisible persons. Those persons were Albert and I, secret, busy and multiform in other dimensions of time, I raised my eyes and the tenuous nightmare dissolved" (*L,* 28). As Yu prepares to kill Albert, he remarks that the "future already exists" (as in the "block"). In Yu's world it is foreordained that he kill Albert; in another world, Albert kills Yu; in still another, Madden kills them both; in still another, they are all figments of some author's "controlled hallucination"; and so on.

Borges's hypostat holds strikingly true to Everett's cosmos. Actually, Borges has incorporated the concept of indefinitely ramifying time and space into many of his stories and essays. To list a few, we are told that "in the sphere of literature as in others, every act is the culmination of an infinite series of causes and the cause of an infinite series of effects" (*OI*, 11). The "Examination of the work of Herbert Quain" (*F*, 81–87), unlike "The Garden," separates fiction from reality, and the direction is reversed: Quain's work begins with the end, which is determinate, and moves back to all the possible causes. In "The Library," Borges alludes to a potential infinity of ramifications for language: "An n number of possible languages use the same vocabulary; in some of them, the symbol *library* allows the correct definition a *ubiquitous and lasting system of hexagonal galleries*, but *library* is *bread* or *pyramid* or anything else, and those two words which define it have another value" (*L*, 57–58). In "The Other Death" (*A*, 103–11) we learn that to "modify the past is not to modify a single fact; it is to annul the consequences of that fact, which tend to be infinite" (*A*, 109). Finally, in the afterword to *The Book of Sand*, Borges suggests branching readings of his work: "I hope that these hasty notes I have just dictated do not exhaust this book and that its dreams go on branching out in the hospitable imagination of those who now close it" (*BS*, 125).

Borges's branching worlds, like the many-worlds interpretation, is perhaps appropriately illustrated by Zeno, once again. Borges and the reader become lost in the infinitely recursive series of eternally incomplete approximations. The Eleatic image of a multiplicity of universes is packed so tightly that the notion of interstices becomes spurious—the interminable division of countless beings and events in countless times and space; the interminably repeated images created by innumerable mirrors; the labyrinths: the Labyrinth.

Man follows the ways of the Earth,
The Earth follows the ways of
Heaven, Heaven follows the ways of
Tao, Tao follows its own way.
                    —*Lao-Tzu*

Quantum —and Textual— Interconnectedness

■ The infinitely packed image of the universe that concluded the preceding section opens the door to this section: the universe as an elaborate network where each member of the whole (the set

of all sets) is intimately linked to everything else. I refer to "field theory," which Fritjof Capra (1975), a particle physicist and prophet of the holistic view, has in mind when he asserts that the universe is a dynamic web of inseparable energy patterns. Although quantum mechanics and relativity theory are incompatible on many points, in general they share the fundamental premises that (1) the universe is an interconnected whole; dichotomies of space and time, matter and energy, gravity and inertia, become nothing more than different aspects of the same phenomena and (2) there is no such thing as observing this interactive whole from a neutral frame of reference. Necessarily and irrevocably, we are inside the dynamic cosmic web.

Matter and empty space, the full and the void, constituted the dichotomy of atomism from Democritus to Newton. According to general relativity, in contrast, the two poles cannot be separated. Wherever there is a massive body, there will be a gravitational field manifested by a curvature of space around that body; all matter is inseparable from space as independent parts of a single whole. Neither is there an absolute distinction between matter and energy. According to field theory, a material particle such as an electron is merely a small domain of the electrical field, within which the field strength has assumed values of enormous magnitude, indicating that a comparatively large field is concentrated in a very small space. Such an "energy knot," which by no means is clearly delineated against the remaining field, "propagates through . . . space like a water wave across the surface of a lake; there is no such thing as one and the same substance of which the electron consists at all times" (Weyl 1949, 171). The field exists always and everywhere, and a part of it cannot be effectively isolated from the whole. It is the carrier of all material phenomena, the "void" out of which "particles" emerge and fade. In ordinary life, we are not aware of this unity; we divide our surroundings into separate objects and events. This division is practical and necessary, but it is not the fundamental nature of reality, today's physicists tell us.

One of the most intriguing proponents of the interconnectedness thesis has been David Bohm, a maverick physicist and one of the chief opponents of the Copenhagen interpretation. Bohm suggests that the universe consists of what is most adequately termed "unbroken wholeness," which can only be described as "that-which-*is*." Space, time, and matter simple *are:* they are the *unfolding* of that which was *enfolded*, the actualization of a potentiality. The totality of the universe is therefore, and

Chapter Six

paradoxically, self-contained, immanent. Bohm's undifferentiated whole, the *enfolded*, is called the *implicate* order. The actualized world of particulars, the *unfolded*, is the *explicate* order. Bohm illustrates the relationship between the two orders with a tropologue, a metaphor. If we put some viscous fluid such as glycerine in a container, place a drop of insoluble dye close to one edge, and turn the container slowly, the dye is "stretched" out in a circle until it seems to disappear. It is still there, but it is now enfolded, implicate. If we reverse the turn of the container, the drop reappears in its previous form. It has become unfolded, explicate. While the dye was in its enfolded state, it existed, though we could not perceive it. Simply because we do not see something or are not presently conscious of it does not imply its nonexistence. It is there, potentially to become explicate. And, it bears mentioning that what is at a given moment explicate must in this sense imply what remains implicate but could have become explicate (Bohm 1980, 140–71).

The combination of the implicate and the explicate orders is what Bohm terms the "qualitative infinity of nature." Much like Whitehead, he argues that there can be no end to the levels of interconnected networks, from the infinitely great to the infinitely small, so that there is no end to the number of interpretations of the universe from a potentially unlimited number of perspectives. This notion calls into question the "thingness" or "beingness" of what has been actualized. Every "entity," no matter how fundamental it may be, depends for its existence on the particular conditions of the implicate order, which is in turn affected by the mutual interconnectedness of the "entity" or set of "entities" under consideration, which have been, or which are, in the process of actualization into the explicate order.

Each "entity" in nature owes its existence in the explicate order to a balance of opposing fluctuations in the mutually reciprocating background, the implicate order, which is incessantly changing in diverse ways. And this "entity" exercises an influence, therefore offering some contribution, though ever so slight, to the universe as a whole. There is no absolute autonomy, hence the properties of a particular "entity" can be attached to no more than ephemeral definitions and conceptualizations. This "process" metaphysics of the universe offers the image of a potential infinity of factors that determine the properties of a particular "entity" at a given instant. From one perspectival grasp, an "entity" evinces a potential infinity of aspects in common with those it possessed during the previous

instant; if not, it would not preserve any form of identity over time. By the same token, it manifests, at that same instant, a potential infinity of aspects differentiating it from what it was. Given the infinity of factors determining the properties of an "entity," nothing can remain identical with itself. Bohm reminds us that since thus far empirical evidence has provided no mode of being that remains eternally defined in any given way, there is no reason to expect absolute verities in the future. In fact, the assumption of the absolute and final nature of perspectives, concepts, and theories contradicts the very spirit upon which science is predicated, in spite of those wide-eyed somnambulists who proclaim the contrary to be the case.

If no "entity" can be in all respects identical with what it was, but always already something different, then any and all definitions and conceptualizations of that "entity" can be no more than skeletons, the most limited of abstractions, and they are austerely circumscribed by contexts, the contents of which are incapable of providing a totalizing reflection of the potential infinity of other contexts; immanence necessarily prevails. Hence all abstractions, according to Bohm, are at most nothing more than approximations, for we cannot hope to encompass conceptually the qualitative infinity of nature. Science, in this view, cannot lead to error-free knowledge. The arduous task of uncovering errors in previous theories reveals that, in our universe of incessant becoming, new phenomena will always and invariably pop up.

Bohm's metaphysics, as I understand it, also implies that mentation and "physical reality" share comparable algebras at the implicit level. Moreover, the ramification of the interconnectedness of things ultimately leads to consciousness, for

> [w]hen one part [of consciousness] is explicit, a tremendous amount is implicit. As we talk, the words are explicit, but the whole meaning is implicit; . . . This implicate order is common to mind and to matter, so it means that we have much of a parallelism between the two sides. . . . Things which are well defined and explicate have to be seen as special features of the implicate order. . . . This idea of implicate and explicate order obviously involves wholeness, because, in the implicate order everything has its origin in the totality; it is folded into the totality. (Buckley and Peat 1979, 157)

Bohm's notion is considerably less determinate than the argument that mind parallels (mirrors) matter, or the

reductionist thesis that mind functions are nothing but brain functions. It does point toward Berkeley's saying that he did not aim to change things into ideas, but rather ideas into things. Bohm's abstract structural similarity between mind and matter has recently come to be known to a few as the "holographic theory," named after the hologram, a dramatic breakthrough in photography by the use of laser beams.[17] An intriguing feature of laser photography is that there is no one-to-one correspondence between the parts of the photographed object and the parts of the photographic plate. Rather, the interference pattern of each region of the plate is equal to the whole—like the noted Buddhist string of pearls, each of which reflects, and therefore contains, the whole. If the corner of a plate is broken off and illuminated, it will reproduce the entire object, though the smaller the piece, the more vague the reproduced image becomes. This description of the "holographic paradox"—as it has occasionally been termed—is necessary, for it illustrates a theme, discussed above in another context and found throughout Borges's texts, especially "The Aleph": *the part is equal to and reflects the whole.* I refer in particular to the "map paradox," which exists in two forms: (1) a map constructed the size of the country to be mapped so as to include each detail (*CB,* 119), and (2) a map laid out on a flat stretch of the terrain in the countryside to be mapped, and constructed in such fine detail that it reduplicates that country faithfully (*OI,* 45–46; Royce 1901). A possible problem with the first map is merely that of identity. Which is map and which is territory? The second map, which interests us here, must, since it is part of the country to be mapped, represent itself in itself, and that representation must represent itself, and so on. This is the essence of Bohm's intriguing metaphysics: if the part contains and mirrors the whole, then it contains itself within itself, and that part of itself must have a part of it which contains itself—Cantor's sets.

Bohm's hypothesis, I believe it has become evident, also bears on Zeno's paradox of movement. For Bohm, notions of continuous and discontinuous movement as occurring over time have been erroneous. According to the traditional view, which inextricably remains linked to the Eleatic problem, an arrow flying from the archer's bow toward the target requires a certain period of time; each increment might correspond to an almost infinitesimal jerk, and during each increment of time, the arrow was in a certain point in space. Bohm's movement of an object unfolding out of the implicate order, in contrast, is more in tune

with our inductive sense of movement. Suppose we put a drop of dye in the glycerine and turn the stirring mechanism n times. We could then place another drop nearby, and stir n times again, and then repeat the procedure n times. Then we would have a series of enfolded "dots" along the curvature of the turns. After their enfoldment, we then reverse the turns, but this time so rapidly that the individual "dots" merge into one another. We now have what appears to be a continuously "solid" curved object. This, suggests Bohm, is analogous to the movement of immediate perception, because the eye is not sensitive to the dye in lower concentrations. One does not directly see the entire movement.[18]

This consideration indirectly returns us to the problem of time. According to one interpretation of quantum theory, every point of continuous space is the *potential* position of a particle, and every instant of continuous time is the *potential* time of an event. The becoming of events, from this potential, is, in the final analysis, *mind-dependent*. Physical events merely occur "tenselessly in a network of relations of timelike separation" (Grünbaum 1967, 55). Thus quantum theory has discretized or quantized some properties—i.e., a quantum event described as the movement of a particle conceived of as a series of discontinuous jumps—which were considered continuous in classical physics. However, time and space remain unquantized. Zeno denies movement (in space and time) along a series. This, it will be recalled, is comparable to Borges's denial of time construed as a series. But since space and time are not quantized (i.e., serialized, made tenseless), quantum theory is not challenged by the arrow paradox (Grünbaum 1967, 114).

Bohm's abstract model is, properly speaking, also tenseless insofar as movement through space and time can be statically presented on paper or a blackboard as a graph, with time corresponding to the "y" axis and space to the "x" axis. The "flow of time" is represented by a line from the bottom angling toward the top. This packs three dimensions into two, providing a static view of what to immediate experience would have been a continuous flow—the equivalent of Escher's above remark that after the creative process is brought to completion, the result is viewed holistically, in static, timeless fashion. The same can be said of movement in our three-dimensional world seen from a four-dimensional perspective (i.e., the static "block").[19] An example will suffice. To reevoke Everett's branching universe, a particle's trajectory over an exceedingly brief time increment might look

Chapter Six something like figure 11. From point A in time, four possibilities are open after an intermediate temporal increment. Assume these branching universes to be a segment of a flatlander's world. Person A at instant A could be aware of his choices B and C, but DEFG remain in the unforeseen future. The most he can do is proceed a step "forward" to either B or C, and then either to DF or FG. In contrast, from our three-dimensional perspective, we can map out his possibilities from within a timeless framework, DEFG being open to our view in simultaneity with ABC. (Note that this tree diagram is a microcosm of Ts'ui Pên's book, a temporal labyrinth that does not unfold linearly but bifurcates into multiple ramifications.)[20]

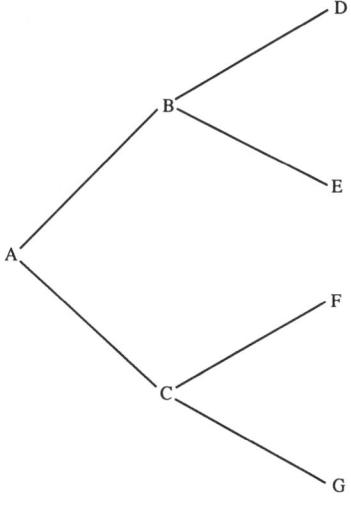

Figure 11

A worthy counterpart to Bohm's "undivided wholeness" in Borges, also prevalent among poststructuralist critics, is the notion of "intertextuality." Each text, within the intertextual fabric, is considered to be a mosaic of citations of other texts; it absorbs them and at the same time transforms them with its coming into existence and with each and every reading.[21] In other words, like the "entities" in Bohm's interconnected whole, intertextuality renders each meaning contingent upon all other meanings, and each meaning is the by-product of family resemblances regarding the commonsense world as it is conceived from a particular culture-laden and language-laden perspective.

Rosalyn Frank and Nancy Vosburg (1977, 582) observe that Borges's notion of "intertextuality" reinforces the "double

reality," the "real" and the "irreal," to be found in his stories. However, this is not the form of intertextuality I have in mind, which is more akin to Gérard Genette's "time of a book": not the limited time of writing, but the limitless serial time of reading (of all readings) and of memory (all memories). In such a "timeless time," along the lines of an "implicate order," Borges tells us that Kafka exercises an influence on Cervantes that is no less important than Cervantes's influence on Kafka. At the outset, this notion reminds one of the poststructuralist thesis, and of comments by diverse writers. For example, André Malraux (1951, 368), for whom each genius that causes a rupture with the past also changes earlier forms. Or T. S. Eliot, also occasionally mentioned by Borges, who writes that "what happens when a new work of art is created is something that happens simultaneously to all the works of art which preceded it. The existing monuments form an ideal order among themselves, which is modified by the introduction of the new (the really new) work of art among them" (1941, 1).[22]

What Is Real?

Obviously Borges's notion is more all-encompassing. It is not only protoactive but retroactive as well. It is simultaneously a radicalization and deconstruction of the well-known fallacies of intention and influence. For example, Borges (*OI*, 106–8), provides a list of unlikely "antecedents" to Kafka's work: Zeno, Han Yu, Kierkegaard, Browning, Leon Bloy, and Lord Dunsany. He then suggests that, though each member of this heterogeneous collection resembles Kafka, they do not resemble each other. Yet, through Kafka, they are interconnected. In other words, Kafka's idiosyncrasies exist in each of those who influenced him, but had Kafka not written what he did, those idiosyncrasies tying each of them together would not have existed. Our reading of Kafka consequently changes, though at times imperceptibly, our reading of Browning, of Kierkegaard, of Cervantes, even of Borges and Shakespeare. Borges concludes that each writer *creates* his precursors. His work modifies our conception of the past, as it will modify the future, since each work and each reading of a work affects, to a greater or lesser degree, the totality of intertextuality in a manner comparable to Bohm's interconnected fabric.

We must confess at this juncture, after brief consideration of Bohm, Wheeler's "tail-chasing dog" model, and Gödel's argument within the "block," events $a \rightarrow b \rightarrow c$ for one observer might well be reversed for another. The notion of intertextual reciprocation regardless of unidirectional time is in this sense not as

outlandish as it might otherwise appear. Interestingly, in "An Examination of the Work of Herbert Quain" we learn that Quain wrote a novel, *April March,* in which events digress from future to past, a regressively bifurcating novel the equivalent of reversing the order of figure 11 and reading it from right to left. Quain, who lays claim in his novel "to the essential features of all games: symmetry, arbitrary rules, tedium" (*F,* 75), evokes, in the prologue to his work, Bradley's inverse world in which "death precedes birth, the scar the wound, and the wound the blow" (ibid.). In this inconceivable Eleatic realm, the eight stories composing another of Quain's works, *Statement,* deliberately frustrate the reader, one of them even insinuating two arguments that lead the reader to think he has invented them. Borges, apparently in all seriousness, reveals that he was "ingenious enough" to extract from yet another of Quain's stories, "The Rose of Yesterday," his own tale of "The Circular Ruins."

Upon reading about Quain's work, we realize that just as a finely detailed map must contain a copy of itself, so Borges wrote a story about an author whose work dupes its readers into thinking they have created the work—all readers are potentially or actually writers, Quain postulated, for readers are now extinct. And one of those deluded readers is precisely Borges, whose "Circular Ruins" implies that very deluding act: the magician was not dreamt by another magician, but possibly by the very son he believed to have interpolated into "reality." Borges is thus textualized, like Quixote before him, as a reader of a work by a character he created. In another manner of speaking, like the map paradox, Borges's "reality" contains his text, which in turn contains Borges and the totality of his "reality."

It is not amiss to assume, in this respect, that Borges's notion of textual interconnectedness actually entails two spheres: (1) the literary, and (2) the conjunction of "reality," fiction, and dream—which in their composite is tantamount to the literary, which at the same time paradoxically remains as a part of "reality." Regarding the first, Borges approvingly paraphrases Paul Valéry, for whom the history of literature "should not be the history of the authors and the accidents of their careers or of the career of their works, but rather the history of the Spirit as the producer or consumer of literature." He then relates Valéry to Shelley, who fifteen years earlier observed that all the poems, past, present, and future, "were episodes or fragments of a single infinite poem, written by all poets on earth" (*OI,* 10). The plurality of authors and works being illusory, a pantheistic supposi-

tion finds "unexpected support," Borges tells us, in the classicists, for whom literature is what is essential, not individual texts or authors. This calls to mind the monism or complete idealism in Tlön, where a book consists of all possible permutations; that is, it is at once One and Many. According to this "holographic model" (i.e., intertextuality) of literature, the part is necessarily a fragment of the whole, and yet it is intricately connected with every other part of the whole: somehow it *is* that whole. And the whole remains the dream of every author: "The practice of literature sometimes fosters the ambition to construct an absolute book, a books of books that includes all the others like a Platonic archetype, an object whose virtue is not lessened by the years" (*OI,* 66).

In this conception of things, the Russellian paradox once again inheres, which bears further on what I have termed the "holographic model." Just as the set of all sets cannot be a member of itself, so the part cannot logically be tantamount to the whole. *Yet it is,* in spite of Russell's interdiction.

Moreover, if to read an author like Kafka is to modify, by establishing interconnections between Kafka and other authors, the past and future of all literature, this act entails a dilemma comparable to the *Meno* paradox. Socrates claims we are all ignorant, so we should argue together in order to arrive, in dreamlike fashion, at the correct opinion. Meno the slave-boy challenges this assumption: we all somehow know without knowing *that* we know. And Plato provides an answer to Meno, invoking a divine source for knowing where to look for knowledge and how the inquirer can know it when she has it. Borges remains to an extent at Meno's side. Nevertheless, his doubts prevail. Is this, he asks, "a legitimate instrument of inquiry or merely a bad habit?" (*OI,* 114). The inquirer, for Borges, and contrary to Plato, is immanent, yet somehow she can know. Such "knowledge" cannot be purely of "reality," but, in Heisenberg's words, it is "knowledge" of our interaction *with* "reality."

This brings us to the conjunction of "reality," fiction, and dream with respect to the fabric of intertextuality. I have alluded to Borges's "The Dream of Coleridge," in which a thirteenth-century emperor dreams a palace and builds it, and a nineteenth-century poet dreams a poem about the same palace, unaware that the structure was derived from a dream. This puzzle gives rise, Borges conjectures, to the notion that the series of dreams, poems, and labors has not ended. Perhaps, in fact, the series is endless, or perhaps the last person to dream

will have the key. At any rate, whoever might have compared palace to poem "would have seen that they were essentially the same" (*OI,* 17).

This is, indeed, a vision of the "qualitative infinity" of a timeless order of textuality accessible solely to what Borges calls a "superhuman performer." Over the long haul, intertextuality, as I have described it, evinces the unity of a terminus ad quem, a final goal. But, of course, that goal is undefinable, a receding horizon in the sense of Bohm's game of science. The interconnected fabric, whether supposedly in reference to "reality," fiction, or dream, necessarily includes both the world of readers and authors as well as the world of texts. After presenting the map paradox, Borges asks why this should disconcert us, or why it should disquiet us to realize that Don Quixote is a reader of the *Quixote,* or Hamlet a spectator of *Hamlet.* He believes he has found the answer. These inversions "suggest that if the characters in a story can be readers or spectators, then we, their readers or spectators, can be fictitious. In 1833 Carlyle observed that universal history is an infinite sacred book that all men write and read and try to understand, and in which they too are written" (*OI,* 46).

Authors, readers, "reality," fictions, and dreams are incompatible according to common parlance. Yet, *enfolded* into the whole of intertextuality, they can be interchanged and *unfolded,* i.e., read, at propitious, or perhaps not so providential, moments. And how can this whole of intertextuality ultimately be defined? Perhaps, as Bohm tells us of the "qualitative infinity of nature," it is simply *that-which-is.* In this respect, Borges, after tracing the progress of the Buddha throughout history from somebody to nobody, concludes that to be one thing "is inexorably not to be all other things. The confused intuition of that truth induced men to imagine that not being is more than being something and that, somehow, not to be is to be everything" (*OI,* 148). Commensurately, according to Bohm's "qualitative infinity of nature" as well as Everett's "many-worlds interpretation," not to be, or to be nothing, is not tantamount to being everything. Following Bohm, it may be said that "reality" is no-thing. It is the annealment of the *implicate* and the *explicate,* neither determinable nor knowable, yet it is not the totality of all things, i.e., it is not to be identified with everything or all things determinable (Bohm 1980, 60).

In the "many-worlds" conception, the universe as a monolithic superposed wave function is, it might be said, though it is not so formulated by Everett but by Wheeler, both everything

and nothing—everything, for it represents the universe's "bootstrap operation" from the time of the primal cosmic glob, and nothing, for, as a wave function, it is not, in everyday language use, *anything*. In the ever-changing flux of Bohm's interconnected universe, "totality" or "reality" are orders with a certain implied thought content; both are incomprehensible, yet they are not merely meaningless. The words "totality" and "reality" inexorably imply for us, locked into our Western culture as we are, something fixed and permanent. But if so construed, a fallacy has been committed. "Reality" as a whole cannot properly be regarded as potentially the content of thought, for thought is a part of that selfsame "reality." In Bohm's (1980, 60–61) non-Aristotelian formulation, the All is thought coupled with that which is not thought—the two "merge and flow into each other, in a single unbroken process in which they become ultimately one"—and at the same time the All is neither thought nor not thought—for the ultimate ground can be neither specified nor known. There can no more be an ultimate form of such thought than there can be the ultimate thought of the book, referred to above by Carlyle, that is written and read and in turn contains its authors and readers.

The dialectic of everything and nothing, of the "totality" of intertextuality, envisages the "holographic model" as an *enfolded* whole, a minuscule part of which can be *unfolded* at a given moment. Whether speaking of intertextuality or the interconnected, self-contained, qualitatively infinite universe, the "holographic model" leads to the conclusion that

> every entity, however fundamental it may seem, is dependent for its existence on the maintenance of appropriate conditions in its infinite background and substructure [the *implicate* order]. The conditions in the background and substructure, however, must themselves evidently be affected by their mutual interconnections with the entities under consideration. (Bohm 1957, 144)

Very significantly, if we substitute "entity" for "text" in Bohm's statement, it is rendered remarkably compatible with the general notion of intertextuality, that is, with Borges's function of literature.

Indeed, if we conceive of the book as a physical object actualized into a text during a reading—interaction with an observer—the whole of textuality, as *potentia,* must be tantamount to the *implicate* order. Each "entity" (word, sentence,

text) constituting part of this background is necessarily governed by the fluctuations in the *implicate* order of each and every other "entity." A change in a given "entity" brings about a reciprocal change, even though ever so small, in the entire fabric. Unlike a book, the text has no predefined or definable boundaries. The text's status is distinct from that of its author, whose posthumous state does not inhibit the text's dynamic character. The text is not overloaded with its author's history, or with history in general, nor does it suffer limitations due to its author's incapacities. It takes on a life of its own, existing at the limits assigned to it by a particular reading, which must be contextualized, for if not, no boundary can be established. Each successive reading creates for the text a distinct context, and with each new context and reading a slightly to radically distinct text is called up, i.e., actualized, with the simultaneous dissolution of other possible texts into the background, where they remain within the fluctuating balance, exercising their force, be it subversive or benign, on all other actualized or unactualized texts.

Since the whole is incessant flux, a process of becoming, all textual meanings, definitions, and concepts are no more than ephemeral, and as a result of the potential infinity of factors at the underlying level, no text can be self-identical over time. Proportionate with Bohm's "qualitative infinity of nature," there can be no absolute and final state for the field of intertextuality, nor can there be, at a given point in time, final knowledge of any static Saussurean semiological slice out of the intertextual whole. All is incessantly changing.

Interestingly enough, the conjunction of the Library and Ts'ui Pên's labyrinthine book is a conceptual hypostat modeling intertextuality as if it were actualized in its totality—an impossible dream, of course, but a fruitful fiction. The Library contains all possible combinations of a finite repertoire of signs, and Ts'ui Pên's book entails all possible bifurcations in time derived from alternative worlds. In addition, from "Pierre Menard" we become aware of the nonidentity of all texts and their contextualized readings. The fragments of *Don Quixote* that Pierre Menard was able to write before he died are, the narrator tells us, a great enrichment of the original. They are the product of creative endeavors not of a Golden-Age Spaniard but of a twentieth-century Frenchman ignorant of the time of which he wrote. The thrust of "Pierre Menard" is, we have been told, rather than writing, the process of reading, for Menard

has enriched, by means of a new technique, the halting and rudimentary art of reading: this new technique is that of the deliberate anachronism and the erroneous attribution. The technique whose applications are infinite, prompts us to go through the *Odyssey* as if it were posterior to the *Aenid*. . . . This technique fills the most placid works with adventure. (*L,* 44)

According to Borges's concept of intertextuality, to read a book written in the past is in a sense to read through the time that transpired from the day it was written to the present. Today, "Martín Fierro" is not merely the epic poem written by José Hernández; it is the one read by Leopoldo Lugones, Enrique Martínez Estrada, Borges, and many others, and it is the poem transformed into Borges's "The End" (*F,* 159–62), which narrates the account of a knife fight between the Negro Recabarren and Martín Fierro, taking up where the poem left off (Alifano 1984, 33).[23]

Hence, to read a fiction is to reread a reading rather than a writing. Borges's writing *is* his own reading, his reading *is* his world, and his world *is* his text, which is first read and then rewritten, to be reread (M. del Río 1978). Menard's fragments of *Don Quixote* are the fragments of a world, which, in a changed context, become a different text, a different world, and with successive changes of context the world is repeatedly altered. A plurality of worlds potentially exist; they are made to be remade, much in the sense of Goodman, Borges seems to imply. In other words, there is, potentially, an infinity of *Don Quixote*s over time, each written by different authors and read by different readers.

The fact is that, to this point in time, no meaning, interpretation, or hypothesis has remained eternally constituted in any given way. Rather, they have all been found, sooner or later and under suitable conditions, to suffer alterations, even in their most basic qualities, and they subsequently transmute into something other than what they were. This being the case, there is no guarantee that any fixed and absolute meanings, interpretations, and hypotheses will be forthcoming. In this manner, Menard's and Cervantes's passages as books are identical. As texts, however, we are told that Cervantes, a child of his times, merely created a mirror-image of his cultural milieu. Menard, in contrast, like a good Nietzschean, knew how to lie, to create artifice, to say what was not as if it were. This is because the writing and readings of his text were properly contextualized. But over the long haul,

Chapter Six

Menard was no more the author of his discourse than was Cervantes; both were written by their respective texts, they were written into their texts. Textuality exercised its authoritarian rule over the feeble prescriptions of mere mortals.

All told, and commensurate with the interconnectedness thesis, to comprehend Borges's texts according to their well-structured architectonic is not simply to grasp the proper meanings of individual words but instead to attend to the dynamic interrelationships set up among the juxtaposed portions of the work. Christ (1969, 113) describes how Borges is like op art; our response is not due to the evocation of a particular emotion—joy, pity, sorrow. It depends on the relations woven into the texture of the story. And, with respect to intertextuality in general, the same inheres. Borges sees literature as past, present, and future, as a dynamic interaction in constant flux. It is "something living and growing. I think of the world's literature as a kind of forest, I mean it's tangled and it entangles us but it's growing . . . it's a living labyrinth" (Burgin 1968, 38).

Intertextuality, I would submit, is by no means exclusively a literary phenomenon. Of course, most proponents of intertextuality extend the notion to historical, philosophical, religious, and other texts. I am referring, in contrast, to one of the hardest of disciplines: physical science. A study by Enrico Bellone (1980) adequately illustrates this point. A historian of science, Bellone offers a "non-Romantic" approach to scientific creativity and texts. There are, he assumes, no solitary geniuses lying under the apple tree awaiting blinding flashes of insight to shake them from their slumber. From the time of Galileo, theoretical physics has always retained a complex network of connections that can be experimentally tested, and they are by and large compatible with the general temper of their times. Theoretical scientists, especially if they discover a rift in the structure of the prevailing scientific cosmology and break out, must translate, by means of their "scientific dictionary," the body of theory, empirical observations, and experimental data from one conceptual framework to another, from one "language" to another.

Bellone offers the case of James Clerk Maxwell, who made use of rules that could establish a correspondence between Michael Faraday's experimental corpus of work and mathematical language already available. Maxwell was a mathematician, while Faraday was not, but he was an experimentalist. Hence they spoke different "languages." To translate one "language" into

another would require a particular "dictionary." Bellone asks us to assume that we are familiar with Maxwell's "dictionary." It is then compared with Faraday's electromagnetic field theory. The first problem is that Faraday's theory implies interaction between electromagnetic and gravitational fields. In Maxwell's formulation, such an interaction engenders paradoxes with no apparent solution. A limited and unfaithful translator would usually omit that part of the text that breeds anomalies and by a fudge factor translate the rest and patch it up so as to make it appear complete. A competent translator such as Maxwell, in contrast, must be in possession of a more extensive "dictionary." Researching Maxwell's background and analyzing his texts, Bellone discovered that Maxwell's "dictionary" comprised an aggregate of theories: Lagrange's and Hamilton's mechanics, the body of electric and magnetic theories, several branches of mathematics, some astrophysics, and more. Maxwell had studied logic, he had an interest in the history of physics, there were philosophical remarks scattered throughout his manuscripts, and he even evoked theological texts on occasion.

Maxwell's "dictionary," comparable in nature to that of any competent scientist, humanist, or writer, is not only vast, complex, and interwoven with intricate levels of correlation, it is also unstable and subject to changes over time. The importance of this open process cannot be overstressed. Like intertextuality, a modification in one area of a "dictionary" and its respective texts will cause slight to radical changes in other areas. And a new theory proposed by a member of the scientific community will bring about greater or lesser shifts in the "dictionaries" of all other members of the community. Bellone (1980, 16) insists that any historian "who sought to uncover the structure . . . of Maxwell's science by using a label as well known and as much abused as mechanism would certainly get lost in myth."[24] By the same token, a scholar desirous of discovering univocal meaning in a literary text on the basis of the author's history and cultural milieu would contribute to a mythification of the text, for, in dynamic interaction with all texts and their readers, there can be no legitimate halting point.

If the jungle of intertextuality affords us no ultimate harmony, neither does nature, which has for centuries been the object of humankind's quest for order, balance, and symmetry. That dream appears to have been ill-founded, as we shall now observe.

Chapter Six *As far as I can see all a priori statements in physics are based on symmetry.*
— Herman Weyl

[I]f we were positively sure the universe is a labyrinth, we would feel secure. But it may not be a labyrinth. In the labyrinth there is a center. . . . However, we don't know if the universe has a center; perhaps it doesn't. Consequently, it is possible that the universe is not a labyrinth but simply chaos, and if that is so, we are indeed lost.
— Jorge Luis Borges

Symmetries, Mirrors, Broken Symmetries

5 Few phenomena have fascinated thinkers more than symmetries, and Borges, for whom "reality favors symmetry" (*F*, 169) and destiny "takes pleasure in repetition, variants, symmetries" (*DT*, 36), is no exception. Borges is, in a very real sense, always on the lookout for symmetries, as if they might reveal some secret plan (Burgin 1968, 125–26). That symmetries have long afforded archetypal security is no more evident than in the writings of Carl Jung and Mircea Eliade. Symmetries give assurance that humankind and its world are in some mysterious manner the duplication of an eternal model. Understandably, symmetry has in a sense meant the proportioned, balanced, and harmonious. This definition pertains not only to space but also to auditory and sensory qualities in poetry and music (G. & D. Birkoff 1932). In addition, symmetry is revealed in the cyclical time of indefinitely repeated identical events among ancient peoples and in the classical Newtonian universe of infinitely repeated fluctuations, which never ceased to give Nietzsche headaches. Time symmetry, where $-t$ and t can be substituted in equations without changing the system, has also been important in contemporary physics until recently. Besides time symmetry, the two general classes of symmetry addressed in this chapter are what I shall call rotational symmetry, whose mirror image is identical, and bilateral symmetry, or the symmetry of left and right, whose mirror images (enantiomorphs) are an inversion (see Weyl 1952).

Regarding rotational symmetry, consider the hexagon in figure 12. Its left side is identical to its right side, and if a mirror

is placed before it, it is faithfully reduplicated. In addition, it can be rotated, and if no value is attached to its six corners, after each sixty-degree rotation it is exactly like it was. The hexagon is the geometrical form used by Borges in constructing his Library. Agheana (1984, 181–83) mentions that the triangle or pentagon would be unacceptable for the Library, because, seen from within, one side of them would not be mirrored by the other. The form must be symmetrical. It is merely a function of idiosyncrasy, however, Agheana continues, whether the octagon or the hexagon is used, since both are equally symmetrical. I would suggest that the choice is far from arbitrary, however. It is not mere coincidence that the hexagonal pattern is perhaps the most frequently occurring ornamental design in human cultures throughout history. A prime example of this pattern in nature is realized in the honeycomb of the beehive. A colony of worker bees, which are of nearly identical size, build their cells around themselves by gyrating about a point while secreting wax in a semifluid state. The form of the cell, one might assume, would be circular, but the capillary action of the wax adhering to itself causes it to cluster at six equidistant points along the circumference to form a hexagon, which is naturally the densest and most economical packing of the parallel cylinders.

What Is Real?

Figure 12

Chapter Six

If a layer of marbles—spheres, the form so dear to Borges—are placed in an inclined tray, they will arrange themselves in a triangular pattern (see figure 12). Three marbles (or circles on the plane) form a triangle and seven a hexagon. If we stack successive layers of marbles upon the first, it is not difficult to visualize the accumulation of these individual cells becoming spherelike, and to comprehend why the narrator tells us that the Library is a *"sphere whose exact center is any one of its hexagons and whose circumference is inaccessible"* (L, 52). If there is a central hexagon, its center must be the center of a central sphere, which in turn will mirror and duplicate the entire sphere in good Cantorian fashion. The hexagon, then, is the imperfect facsimile of divine perfection. This imperfection, in the Library, profanes the sacred, taints the pure.

Interestingly enough, the hexagonal structure of each minuscule gallery in Borges's Library reduplicated indefinitely such that the Library appears to become more and more spherelike as the sides of each hexagon become smaller relative to the whole (i.e., the fictive approach to the limit or infinity used in calculus) creates an image of that rage in recent years going by the name of *fractal geometry*. Spawned by Benoit Mandelbrot, it is the dream of some hopeful physicists that fractals may be capable of orderly patterning disorder, or chaos. Traditional geometry treats regular forms: straight lines, smooth curves, shapes with perfect symmetry. Yet nature rarely displays such harmony. Rather, edges are ragged, surfaces at a closer look become tangled networks, and, regarding bilateral symmetry, the left side is never a faithful inversion of the right side. As Mandelbrot puts it in the opening lines of *The Fractal Geometry of Nature* (1982), "Clouds are not spheres, mountains are not cones."

A common example of the generation of fractals is the "Koch curve" consisting of the infinite reiteration of a recursive operation (see figure 13). Starting with an equilateral triangle, the first step yields three identical triangles on each side, making a six-pointed star, the second gives twelve more triangles, and the third an additional thirty-six to produce a "snowflake" image. At the theoretical upper limit we would have a continuous "curve" with an infinite number of infinitesimal barbs or excursions. Like Borges's Library, the "curve" would be monstrous. It would be impossible to visualize. Possessing no tangent, it would change direction radically at each and every point. In a sense, it would be infinitely irregular, or infinitely

regular, take your pick—quite understandably, at the outset mathematicians tended to recoil in horror when faced with Mandelbrot's unruly fractals. In the sense of a recursively generated pattern, the "Koch curve" would create an image comparable to Borges's Library seen from afar. However, the Library's inhabitants are not as fortunate as we are, with our ability to take in the two-dimensional pattern in one perceptual gulp. Inextricably caught within the Library's confines, they are capable of perceiving no order, though the desperate hope remains that somehow the Library must evince some sort of order.

Figure 13

With respect to bilateral symmetry, classical physics demonstrated nothing indicating any inherent difference between left and right. It was supposed that just as all points in space are equivalent, so are the two sides of a bilaterally symmetrical object—Tweedledum and Tweedledee. Since antiquity, however, mythical thinking has taken sides with one pole or the other of a struggle of dichotomies, for just as good is opposed to evil, so in myth there needs be a value-laden opposition in all things. Thus the one side becomes *sinister* (from L. *sinistrum,* left) and the other *right* (dexterous). Such mythical thinking has tended to pervade natural language in its description of the world. We have, for example, *negative* and *positive* charges, *negative* and *positive* numbers, and *levarotary* and *dextrarotary* sugar molecules. Nevertheless, the natural numbers, it is generally presumed, remain absolutely symmetrical, with zero at the center, as if a mirror had been placed at the central point.

Bilateral reflective symmetries also allure Borges. In "Three Versions of Judas," The Word, when made flesh, "passed from ubiquity to space, eternity to history, from limitless satisfaction to change and death" (*L,* 96). This lowering of The Word to a mortal condition is a transformation into its opposite, a sacrifice that demands a symmetrical sacrifice of a man, who, rather than the promise of everlasting life, must be condemned to

the "perpetual fires of Hell." Judas was precisely that man, destined to the lower order, which "is a mirror of the higher; the forms of earth correspond to the forms of Heaven; the spots on one's skin are a chart of the incorruptible constellations, Judas in some way reflects Jesus" (*L*, 96-97). Of course, such reflection implies bilateral mirror imagery rather than repetitive rotational symmetry. Rodríguez Monegal (1978, 31 & 348-49) writes of Borges's early fear of mirrors and how he always harbored a hatred for them. Perhaps the infinity of mirror reflections disturbed Borges not so much because it multiplies and simplifies reality but because it generates endlessly not the same but an inverted, and not-exactly-faithful image. Judas "in some way" reflects Jesus, yet the inversion, especially given its slight distortion, inevitably produces tension. Such tension is also revealed in Borges's comment on Don Quixote and Sancho Panza as a "symmetrical and persistent disharmony" (*OI,* 141). There is no mere reflective opposition here but an incessant struggle as a result of the imbalance of opposites.

In "Theme of the Traitor and the Hero" (*L*, 72-75), it is discovered that Kilpatrick has betrayed Ireland. The problem is that he was worshipped by his compatriots as a hero and the greatest hope for their movement. As a solution to this dilemma, he agrees to sign his own death warrant, knowing full well that his fellow conspirators will kill him anyway, but his execution is carried out in the guise of an assassination. Thus he is lost to the cause, yet his redemption as a savior figure is salvaged. He is both hero and traitor and neither one nor the other at the same time. The left side becomes the right in its mirror reduplication, but the right carries with it vestiges of the left, which has possessed, from the beginning, a remnant from its counterpart—a gravitation toward polarity with one side privileged over the other rather than pure mirror symmetry. Similarly, in "The Shape of the Sword" (*L*, 67-71) John Vincent Moon betrayed his friend, who had once helped him. The man telling Borges the story of this wretched individual confesses at the end that he is Moon. He is at once himself and his other, a respectable man and a coward. He reflects himself but can be no more than half of the composite of the two images. He is destined to disclose either one or the other of his half-selves, for each half contains the seeds of the other; there is no all-or-nothing distinction between them, like the *Yin* and *Yang*, both of which contain within themselves a bit of the other.

The German warrior Droctulft of "The Warrior and The Captive" (*L,* 127-31), who left barbarism for the culture of the

city, is juxtaposed with a captive English girl who had renounced civilization for life with the Indians. Though 1,300 years and an ocean separate the two, their story is one and the same, the "obverse and the reverse" of the coin (*L,* 131). In contrast, a sequence of events and its opposite are embodied within the same person in "Emma Zunz." The story she fabricated of killing Lowenthal to avenge her father was true, but only symmetrically true: it was its own inversion.

What Is Real?

Symmetry of metaphysical doctrines, theologies, and ideologies are also to be found in Borges, such as the pantheist "who declares that the plurality of authors is illusory," and the classicist for whom "literature is the essential thing, not the individuals" (*OI,* 12). Or, as discussed in Chapter One, nominalism and realism, idealism and materialism, converge in the totality of Borges's writing to dismantle traditional boundaries, to allow for flow, successive interaction, incessant change.

The most imposing construction of bilateral symmetries as opposed to asymmetries, however, is found in "Death and the Compass," which Ernesto Sábato (1945, 104) calls an "extreme case of geometrization and the legitimate descendent of the scientific novel inaugurated by Poe." The interplay of three (triangle) and four (parallelogram) in this story is common knowledge among Borges scholars and needs little further comment. Suffice it to recap that on the third day of December, the third Talmudic congress began in which Yarmolinsky, victim of the first murder, who had endured three years of war in the Carpathians and three thousand years of oppression, was to be a participant. Yarmolinsky's murderer left a note in the typewriter, "The first letter of the Name has been uttered," along with Hebrew books on the Tetragrammaton and other esoteric topics. This homocide, occurring to the north, was followed by a second, to the west of the city on January 3, and a third to the east on February 3, both also accompanied by notes that the second, and third and final, letters of the Name had been uttered. However, since the obvious Name, JVWV, contains four letters—though three types of letter—Lönnrot concluded that there was surely to be an additional homicide somewhere. To his surprise, on March 1 Inspector Treviranus received a large envelope containing a letter signed "Baruch Spinoza"—that most architectonic of philosophers—and a map of the city with three lines drawn in red ink between the site of the three crimes. The letter prophesied that on March 3 there would not be a fourth murder. Treviranus sent the map and letter to Lönnrot, the mental geometer. After studying them, Lönnrot noticed

Chapter Six   that the events called for "symmetry" both in time and space. The locations of the three crimes mapped out an equilateral triangle, which was asymmetrical and hence inadequate: for time to be "symmetrical," there could be no murder on March 3, since February had three fewer days than December and January. With calipers and a compass, Lönnrot "completed his quick intuition," pinpointing the spot on the map where the fourth crime must occur, on the March 6. He then left for Triste-le-Roy to become Scharlach's victim.

The note Treviranus received was correct: there are three types of letters in the Name, but one letter is repeated, giving four actual letters. The interplay between three and four, and the generational principle whereby the triangle, by duplicating itself, becomes a quaternary structure, is effectively elucidated by Robert Carroll (1979, 325–26) and related to Bruno and Renaissance philosophy. Another aspect of the three-four interaction, however, has been overlooked by most Borges critics. Not only does trinary space become quaternary, the same occurs to time also, but not as Lönnrot had calculated. The first three murders occurred on the third day of three successive months, and each murder was separated by thirty-one days from the others, to give equidistant temporal separation. The fourth murder, Lönnrot obviously inferred, must occur on the sixth of March to separate it by thirty-one days from the third event. Lönnrot rushes to Triste-le-Roy precisely on the sixth, but after being disarmed there, Scharlach tells him: "You are very kind. You have saved us a night and a day" (*L,* 84). Why is this? Scharlach explains that Lönnrot had omitted the fact that actually the deaths had occurred on the fourth days of the first three months, for the Hebrews compute days from sunset to sunset. Lönnrot had been calculating with threes, which were fours in Scharlach's scheme, hence the fourth death was to occur on the seventh rather than the sixth day of the month. Very significantly, then, Lönnrot's scheme, developing linearly through time, was viewed from Scharlach's scheme *in toto.* Moreover, the symmetry of time Borges speaks of is not time in the sense of reversible time. Time, in Scharlach's conception of things, is metaphorically another dimension of space, roughly comparable to the space-time continuum.

Classical physics depicted a linear flowing concept of time, like that of Lönnrot. Special relativity changed that picture entirely. It is now conceived that there is no *same moment* for everything. An instant is identified with a single point in space,

and the totality can be conceived only in "block," or from Scharlach's view, when tracing Lönnrot's path through his labyrinth. Time, in this relativist conception, is symmetrical: retrogressing is no different than proceeding forward. This, of course, violates our normal perception of time, that is, psychological time, which accords with the classical entropy principle predicated upon time asymmetry. If an egg falls off the wall and splatters on the ground, any formula the physicist devises can hardly, we intuitively suppose, reverse all the molecules of that event and bring the egg back together—even though no basic laws would be violated if this happened to occur.

Yet, to the question, "Where, actually, does time asymmetry come from?" a definite response is not forthcoming. Moreover, when focusing on the quantum domain, the origin of time asymmetry becomes even more baffling. At this level, the collision between any two particles appears to be completely reversible: no preference for past or future can be noted (Davies 1974, ch. 1).

At the quantum level, the notion of spatial symmetry once seemed to be incontestable. A host of mirrorlike symmetries, especially parity symmetry—that positive and negative charged particles can be interchanged—and the "eight-fold way" of quark symmetries, led some to believe that, with the possibility of such apparent harmony, the ultimate building blocks of nature were to be found just around the corner. Unfortunately, it was soon discovered that in radioactive decay, elementary particles known as neutral K mesons violate such mirror symmetries; instead of a neutral right/left-handed symmetry, nature showed a bias for one type of handedness. Specifically, it was discovered that the radioactive isotope cobalt-60 emits more radioactive particles from the south pole of the nucleus than from the north pole (Lee 1966).[25] This proved to be disappointing for many physicists, especially those, such as Wigner, whose penchant for symmetries was not easily discouraged. Wigner (1965) has located the origin of the temporal asymmetry of quantum measurements in the consciousness of the observer, a "psychophysical parallelism" tying mind to nature. If such is indeed the case—which most physicists deny—it might conceivably account for the breakdown in mirror symmetries. The fact remains, however, that the desired harmony does not at this point seem to be forthcoming.

Asymmetry might exist at the very origin of life itself. All amino acids in living matter have the same left-handed twist,

Chapter Six which accounts for the uniformity in the handedness of all DNA helices. Did this originally occur by sheer accident? Could the twist as easily have been right-handed? The truth is that asymmetry tends to breed asymmetry, as Hofstadter (1979, 130–40) demonstrates in his disquisition on recursive series. Ordinary crystals are built up by the fastidious repetition of the same structure in three directions over and over again. If, on the other hand, after a cycle is completed, the end point does not coincide exactly with the starting point, the subsequent repetition will not be identical, and a spiral rather than successive cycles will result. Schrödinger (1967, 64–65), some years before the connection between genes and DNA was known, predicted that genetic information was stored in such "aperiodic crystals," which can grow into an almost—but not quite—symmetrical structure. Once their aperiodicity is established, there is virtually no limit to the size of the aggregate that can be generated.[26]

Aperiodicity, like asymmetry, qualifies human mental constructs as well as the world in general. The Gödelian sentence invariably found in sufficiently rich and complex formal systems, Heisenberg's uncertainty, the incompleteness of all descriptions, linguistic vagueness and ambiguity, all testify to a universe that is inextricably tainted somewhere, and at some time or other. All elaborate intellectual constructs conceal a flaw, if no more than of the most minute variety. Gilbert Chesterton, in *Orthodoxy,* pinpoints the dilemma:

> The real trouble with this world of ours is not that it is an unreasonable world, nor even that it is a reasonable one. The commonest kind of trouble is that it is nearly reasonable, but not quite. . . . It looks just a little more mathematical and regular than it is; its exactitude is obvious, but its inexactitude is hidden; its wildness lies in wait." (quoted in Gardner 1979, 211)

Although Borges makes a case for symmetry, his inclinations ultimately point toward asymmetry, toward novelty rather than sameness, uniqueness rather than mere repetition. Indeed, there is an omniscient tension in Borges between symmetry and asymmetry. Symmetry signifies rest and binding; asymmetry motion and loosing. We desire time to be cumulative, as Borges subtly and ironically demonstrates: the "idea of progress," so dear to the Western mind, cannot easily be eradicated. Yet we also long for the timeless. Thus the tension between symmetrical models of simplicity, which are inexorably over the long haul

a myth, and the labyrinth, which is a universe of chance, chaos, the multiplicity of the unknown (Bunge 1963).

This tension surfaces in "Pierre Menard, Author of Quixote." The reader is provided with a list of Menard's writings prior to his attempt to write the *Quixote*, which is revealing. On the one hand, there are works predicating the unique. These include a symbolist sonnet that appeared twice *with variants*, a monograph "on the possibility of constructing a poetic vocabulary of concepts which would not be synonyms or periphrases of whose which make up our everyday language" (*L*, 37), a treatise on different solutions to the problem of Achilles and the Tortoise, and an invective against Valéry, which is the "exact opposite" of his true opinion toward the French poet. On the other hand, there is a monograph on "certain connections and affinities" between Descartes, Leibniz, and John Wilkins, another on Leibniz's *Characteristica Universalis*, a rejection of the possibility of innovating the rules of chess, a worksheet on George Boole's symbolic logic, and an examination of "the essential metric laws of French prose."

The first grouping, entailing a penchant for difference, change, novelty, is juxtaposed with the second, imaging order, harmony, symmetry, repetition. This tension is magnified subsequently when we are told that Menard was inspired by two texts "of unequal value": one, Novalis's philological fragments, "which outline the theme of a *total* identification with a given author," and the other, "one of those parasitic books which situate Christ on a boulevard, Hamlet on La Cannebière or Don Quixote on Wall Street" (*L*, 39). Menard abhorred the second, which is fit only "to enthrall us with the elementary idea that all epochs are the same or are different" (*L*, 39). Significantly, he did not want to compose merely another *Quixote,* he desired to pen *the Quixote itself,* identical to the early seventeenth-century work down to its most minute detail. After an initial abortive attempt, his ambition is realized in a collection of fragmentary passages. But the project backfires; his critics fancy that his text has become, rather than identical to the *Quixote*, a set of successive differentiations that are even contradictory at some points. It is appropriate that Menard's *Quixote* remain unfinished. The fragments, like all works, the collection of which constitutes the fabric of intertextuality, will remain forever open. Intertextuality, of course, is incompletable, that is, asymmetrical, for the finite reader. A cosmic perspective of the symmetrical whole of "intertextuality," if

Chapter Six   indeed there be any, is not for us. We can do no more than read the text through a small window that affords us an action-packed movie of constant flow; things happen in the "now" because we perceive that time transpires: the "now," wherever it is, wherever we are, at some point in the labyrinth. Borges's disposition to oscillate between symmetry and asymmetry reveals a most penetrating insight.

Chapter Seven **Suspended within Language**

The realist-nominalist controversy with which I initiated this inquiry aside, Borges, in his metaphysical fiction and his essays, seems to adopt a form of linguistic idealism according to which we are suspended within the confines of language, unable to step outside the limitations with which we must operate. This idea has been on the increase since Nietzsche, for whom our very thinking must come to a full stop once we refuse to bow to our linguistic constraints. Martin Heidegger (1971, 134), picking up on the same theme, tells us that we remain committed to language and "can never step out of it and look at it from somewhere else. Thus we always see the nature of language only to the extent that language itself has us in view, has appropriated us to itself." More recently, Derrida's (1974a, 158) notorious assertion, "*There is nothing outside the text*" reiterates the premonition that language does not refer to "reality": the very word itself is inside language, as incapable of escaping the conceptual fabric as any other word.

For every language there is a frontier, the crossing of which is strictly prohibited by that language's implicit set of injunctions, yet the frontier is frequently crossed. Such an act of crossing entails a paradox. On the one hand, language's very existence is predicated on the impossibility of saying everything

Chapter Seven

(the existence of grammar implies the existence of nongrammar, consequently, there is an implied exclusion: "This cannot be said"). On the other hand, there is *nothing* that cannot be said—that is, nothing need necessarily be left unsaid, for there is hardly a rule one does not eventually break, either intentionally or unknowingly. In another manner of speaking, there are no explicit rules for what can be said outside the rules (i.e., you cannot speak *about* what cannot be said, at most you might be able to say *it,* otherwise you are consigned to silence; but if it is said, then some implied rule, or nonrule as it were, has been broken). Borges is, as to an extent we all are, trapped within such a quandary: to describe the ineffable Aleph, Tzinacán's effort in "The God's Script" to narrate his mystical experience, to discover the key to Ts'ui Pên's labyrinthine book, Averroes's problem in understanding Aristotle's *Poetics,* whether or not the Library possesses any sort of order, and so on.

During the painful birth of quantum theory, language also caught some physicists red-handed. They thought they would pilfer a few words here and there, put them in one-to-one correspondence with the mathematical formulations they were using so beautifully and parsimoniously to describe their new "reality," and everything would be fine. But, to their dismay, language refused to be pinned down. This revealed nothing new about language; it did serve to disclose a long-cherished but deluded hope that an intelligibly constructed set of signs could eventually be made to correspond to nature.

In this, the final chapter, following brief commentary on the problem of couching relativity and quantum theory in natural language, I introduce the controversial "incommensurability thesis," according to which logical lines of comparison and contrast across scientific "paradigms" or radically distinct cultures is well-nigh impossible. I then extend the discussion of three stories by Borges that have a bearing on these issues: (1) *textual interpretation* across "incommensurable" cultural forms of life and languages which may be partly intertranslatable ("Averroes' Search"), (2) *nonlinguistic "textual interpretation"* in the absence of two different languages which share certain commonalities, whether determinable or indeterminable ("The Immortal"), and (3) *textual interpretation* of identical texts with radically distinct *contextualizations* ("Pierre Menard"). I do not presume this chapter to be a detailed exposition of the incommensurability thesis, nor is it a critique of hermeneutics. Rather, I continue my efforts to reveal the relevance of Bor-

ges's texts to a broad spectrum of twentieth-century thought. Ultimately, I trust, the full import of Borges's prose will become apparent, especially insofar as it incorporates the state of conflict and tension, characterized by the spirit of our times, that has arisen out of the duel between, as well as the attempts toward abolishment of, the dichotomies, oppositions, contrasts, and contradictions in traditional Western thought.

This conflict is profound in Borges. His work represents a definite projection toward the postmodern. He abandons all fruitless quests for grounding principles. "Reality," he repeatedly asserts, is made, not found. His texts draw attention to their own fictive devices rather than lean toward the illusion that they are chunks of real-life experience. At the same time, as has become evident throughout this study, Borges's lingering nostalgia places him squarely within the modernist mind-set: his longing for the Absolute perseveres.

Suspended within Language

We cannot begin with complete doubt. We must begin with all the prejudices which we actually have when we enter upon the study of philosophy. These prejudices are not to be dispelled by a maxim, for they are things which it does not occur to us can *be questioned.*
—C. S. Peirce

■ James Jeans's Great Thought, that "mind stuff," it must be emphasized, is purely mathematical. Nature seems more and more the product of a mathematician, and the scientist, on initially creating his mathematical constructs with which to account for the world, increasingly uses mind and mind alone. Both mind and number, working together harmoniously, "transform the world," that is, in the sense that the world is incessantly remade. Indeed, for over fifty years natural science has been on the threshold of a new era epitomized by Warren McCulloch (1965, 2), who, when asked what problem would serve to guide him in his scientific career, replied: "What is a man so made that he can understand number and what is number so made that a man can understand it?"

But which mathematics are we speaking of when we speak of "mathematics"? The question is relevant, for different

Language against Itself

Chapter Seven

mathematics have been used in repeated efforts to describe the world. As I briefly sketched out in Chapter Four, during the early days of quantum theory there were, as there still are, competing views, each with its particular mathematical toolbox. Heisenberg used matrix mechanics and Schrödinger wave mechanics. The problem was that the two descriptions are contradictory. Matrix mechanics does not lend itself to visualizable models; it is purely formal, a description of discontinuity, with no possible account of the whereabouts or nature of a particle in transition from one place to another—a frustrating Cheshire cat act. In contrast, wave mechanics is predicated on continuity, and it is very roughly visualizable. After a period of bickering between the Schrödingerians and Heisenbergians, Dirac finally demonstrated that, by his transformation theory, a wedding could be consummated between matrix and wave mechanics, and the two were thereafter known as the "matrix representation" and the "wave representation." This combination constitutes the essence of Bohr's complementarity principle.

It might be stated that the two mathematical representations of quantum phenomena lend themselves to "intertranslatability" (that is, intertransformability from one state to another) with the precision of a surgeon's scalpel. However, since in concert the two representations refer to incompatible attributes, and since both of them belong to the same general theory, each, by and of itself, is incomplete. Each partly describes the totality of the electron's behavior, with the combination of both providing a more complete, but never a completable, picture (Sklar 1985, 150–55).[1]

During the emergence of quantum theory, physicists also became increasingly aware of another problem: the incompatibility of natural and formal languages. Heisenberg recollects how during the 1920s the world's prominent physicists labored to solve the problems of quantum mechanics, finally to arrive at what the majority believed to be the most acceptable solution. While their initial descriptions were almost exclusively mathematical, during subsequent conversations and debates they inevitably retreated into natural language. This presented a predicament that continued to defy resolution, baffling them at every turn. Heisenberg vividly portrays the collision between customary language use and this new view of reality, which produced an uncanny effect in the physicists' thinking. He goes on to testify, after remarking that this dilemma is a problem of language as much as of physics: "I remember dis-

cussions with Bohr which went through many hours till very late at night and ended almost in despair; and when at the end of the discussion I went alone for a walk in the neighborhood park I repeated to myself again and again the question: Can nature possibly be so absurd as it seemed to us in these atomic experiments?" (Heisenberg 1958a, 175)

Heisenberg explains that Bohr's complementarity principle eventually encouraged physicists "to use an ambiguous rather than an unambiguous language, to use the classical concepts in a somewhat vague manner . . . [but] when this vague and unsystematic use of language leads into difficulties, the physicist has to withdraw into the mathematical scheme and its unambiguous correlation with the experimental facts." This use of language, Heisenberg continues, is in many ways satisfactory, since "it reminds us of a similar use of the language in daily life or in poetry." He then writes of the function of complementarity in much the spirit of Bohr insofar as it is not confined exclusively to the atomic world. It comes into play in everyday life, for example, "when we reflect about a decision and the motives for our decision or when we have the choice between enjoying music and analyzing its structure" (Heisenberg 1958a, 179).

Now, the "experimental facts" Heisenberg speaks of are nothing more than black spots on a photographic plate or water droplets in a cloud chamber. They reveal nothing about subatomic "entities," which cannot be defined exclusively either as things or as waves. Heisenberg gradually became accustomed to considering them not as a form of "reality" known to the senses, but as a kind of "potentia", i.e., the Heisenbergian counterpart to a set of "uncollapsed" superposed waves. By this time, Heisenberg (1958a, 181) continues, the natural language that physicists were using had already begun to "adjust itself, at least to some extent, to this true situation." Such adjustment was demanded of natural language by the formal language describing the physicists' "reality," but this presented a problem: the formal language was not compatible with patterns of traditional logic.

Following Heisenberg—and he is only one among a host of physicists who have written on this "language problem"—it is not in the least radical to propose that scientific theories, the most revolutionary of which ultimately lead to completely new cosmologies, entail world-making of the first order by means of novel language use, whether natural, formal, or both. In the beginning, when a new world-image is slowly being forged,

unexplained terms and sentences for which there are as yet no clear rules are used in vague and ambiguous ways. The most extreme cases even demand what at the outset appear to be illegitimate descriptions that at times even severely contradict our perceived world, e.g., "The earth travels around the sun," "Light travels along a curve," "An electron is both a particle and a wave, and at the same time neither," and so on.

But this is as it should be. For language, whether we know it or not and whether we like it or not, is incessantly used and abused, modified and hammered into new forms in order that it fit the standards of new world-images. Feyerabend (1975, 257) remarks along these lines: "We see . . . how essential it is to learn talking in riddles, and how disastrous an effect the drive for instant clarity must have on our understanding." For instance, classical logic was born only after rhetoric was sufficiently developed so as to serve as a starting point. Arithmetic was developed before a clear understanding of number could exist. Full comprehension of the Copernican universe required a couple of centuries—in fact, the Copernican model was called "Copernicus's paradox" for some generations after its author's death. The idea that a stone does not fall "straight down" when dropped but follows a long trajectory commensurate with the earth's rotation remained a difficult pill to swallow for some time after Galileo. There is as yet no clear-cut consensus on the second law of thermodynamics, and relativity and especially quantum theory continue to remain enigmatic. Were there, during each of these transition periods, a demand for instant clarity, science would have remained virtually paralyzed—and it hardly needs saying that such a demand for everyday language and the arts would be stultifying.

Friedrich Waismann discusses what he terms "clarity neurosis," which threatens to drive the scientist to a fear of vagueness and of speaking in circles, eventually to become tongue-tied, continually asking himself whether what he is doing makes perfectly good sense. "Imagine," Waismann continues, "that the pioneers of science—Kepler, Newton, the discoverers of non-Euclidean geometry, of field physics, the unconscious, matter waves or heaven knows what—imagine them asking themselves this question at every step—this would have been the surest means of sapping any creative power" (Waismann 1959, 359–60).

Obviously Waismann is not proposing that the scientist cease to strive for clarity; rather, it should at times, though not

always, be the primary consideration in his search for answers. Yet an initial effort toward the best possible answer, even though vague, ambiguous, or even paradoxical instead of absolutely clear, can place the scientist in hot water: when he believes he has hit upon a solution, he immediately attempts communicating it with his colleagues, but he is often met with suspicion, mused cynicism, disinterest, or even sheer incredulity. Effective communication of his idea invariably requires some time—in fact, Max Planck once observed that a new theory's acceptance often requires that the older generation first completely die off.

Much the same can be said of the initial response to departures from the beaten path in the arts. "Random music," which is totally indeterminate and hence can hardly be subjected to critical discussion, let alone analysis, seemed to the ears of its early listeners devoid of all significance. It was such a radical break with cultural conventions as to appear completely nonsensical. Yet it eventually compelled its listeners to make explicit their convictions, prejudices, and doubts; only then was it appropriately perceived. In literature, writers like William Burroughs created images by cutting up their texts and recombining them in random or near-random fashion, which was at the outset considered incomprehensible by the reading public accustomed to thinking in purely linear terms. Today's aware readers, in contrast, are familiar with these methods and encounter little to no difficulty absorbing the texts. A sense of timelessness, somewhat reminiscent of the "block" conception of the universe, is created by writers the likes of Beckett, whose novels *Molloy, Malone Dies,* and *The Unnamable* do not allow us the security of knowing whether we are listening to the characters' ruminations for one hour, ten days, or a year. And in painting the strange works of Jackson Pollock are simply *there;* there is no depth, no space, no indication of temporal sequence during the original creation or upon viewing the canvas. What appears to the senses is no more than a static, randomlike present. The disconcerting effects I speak of here are not limited to our times. E. H. Gombrich (1960) effectively demonstrates how, throughout the ages, art has been born of art, painting has altered perception and created tastes, and in the process our "reality" has been transformed by art rather than the other way around. In each of the cases I have rather arbitrarily selected, whether scientific or artistic, a certain period of time was required for the general public, and even the creators of new modes of perception, expression, and thought, to become

Chapter Seven comfortable with the novel world-images that were gradually appearing on the scene.

Now, I have addressed myself—perhaps too briefly, but the point is essential—to two representations of quantum theory, each of which is describable in classical logic, but their combination creates an inconsistency. By the complementarity principle, the two representations supply each other's lack, and their conjunction serves to fill out a larger picture. One might tend to qualify Heisenberg's matrix interpretation and Schrödinger's wave interpretation as incommensurables. But they are not. Though incompatible, recall that at a deep level they delineate one and the same general theory, and they are intertranslatable, hence they are not incommensurable, for if so, there would be no basis for translatability or even for comparisons of the two. The distinction between incompatability and incommensurability is important to the discourse being unfolded here. Two different formal languages describe the incompatible representations in quantum theory, and precise lines of correspondence between them render them intertranslatable. In contrast incommensurability, as I use the term, involves what are considered to be mutually exclusive languages delineating and describing distinct conceptual frameworks. I refer to the incommensurability thesis of the "radical meaning variance" (RMV) theorists, Feyerabend, Hanson, Kuhn, Polanyi, and others.[2]

The incommensurability thesis, a radical article of faith, was in part stimulated by the shock that relativity and quantum theory caused to the portentious belief of nineteenth-century Newtonian physics that, with a little mop-up work here and there, absolute knowledge of the entire universe was close at hand. Briefly, Popper initiated this revision in the philosophy of science. Departing from the school of logical empiricism, he argued in his *Logic of Scientific Discovery* (1959) for a distinction between confirming a scientific theory and refuting it.[3] He proposed that the task of the scientist should be not to verify but to falsify theories. In other words, since verification can go on forever, we shall never know if a theory is true or not, but if we can falsify it, we can at least know that it is *not* true. Since the latter 1950s and the 1960s, the RMV theorists, either drastically expanding Popper's general thesis or working independently of it, have introduced the "language problem." In very general terms they argue that: (1) observations and facts are theory-laden, (2) meanings of theoretical terms are theory-dependent,

Suspended within Language

(3) theory is worldview laden (i.e., bound to a given set of presuppositions), and (4) worldviews are language-dependent. Language by and large organizes our world, whether we are speaking of fiction or fact, verse or verity.

In order to sharpen the distinction between incompatibility and incommensurability, the first, in visual terms, can be roughly compared to perception of the Necker cube. Either the face is up or it is down, but never both in simultaneity. We can compare the two incompatible cubes in our memory, but presumably not while attending to the drawing, for in so doing we immediately actualize one of the two cubes. According to the above discussion of the Necker cube, a fourth-dimensional mathematical time dimension, if Dobbs is indeed correct, is the counterpart to the complementarity principle that is capable of providing an account of both incomplete pictures. Incommensurability, on the other hand, entails a totally different ball game. Figure 14 is used by Feyerabend (1975, 226) to illustrate the concept.[4] We automatically attempt to force the two-dimensional drawing into three-dimensionality, but an anomaly arises. It is neither exclusively three-pronged nor two-pronged; its organization is neither cylindrical nor rectangular. And the cylinder in the middle fades into nothingness as the eye approaches two-pronged, rectangular stimuli. There are no criteria for translating one mode of organization into the other. They simply cannot be compared rationally. So it is, according to the RMV thesis, with the "languages" of incommensurable scientific conceptual frameworks: they compel one to see the world in a particular way that is unavailable to other "languages."

Figure 14

For a "contextual" counterpart to figure 14, which will perhaps be more appropriate for a discussion of RMV incommensurability, consider Marcel Duchamp's putting a urinal on a pedestal in an art show. As with all art insofar as it is perceived *as* art, there is an implicit proposition, "This is art," which remains covert, the most fundamental given for the viewer.

Chapter Seven   When the urinal is placed in an "art 'text,' " the immediate response from a viewer unaccustomed to such bizarre twists might be a reaction to the implicit proposition with the exclamation: "This is *not* art," or in other words, "The proposition, 'This is art' is false." It might appear that we have merely an incompatibility rather than incommensurability. But we must be mindful of the role contextualization plays. The urinal customarily pertains to, say a "biological functions 'text.' " Within that "text," the implicit proposition, "This is a urinal" is no cause for alarm, but when placed in an "art 'text' " with its assumed implicit proposition, there is at least initially another quite shocking situation. The proposition is not simply negated, however, when the outcry, "This is not art" is emanated. Rather, the implication is that this violent recontextualization has generated nonsense, absurdity, or perhaps mere meaninglessness, for as yet there exist no lines of transformation ("intertranslation") between the two "texts." Duchamp had forged an entirely new reading of an item from a context that was at the outset perceived to be incommensurable with an "art 'text.' " In the process "urinal" necessarily took on an entirely new meaning.

In Feyerabend's (1975, 224) words, scientific theories (Kuhnian "paradigms") are "sufficiently general, sufficiently 'deep' and have been developed in sufficiently complex ways to be considered along the same lines as natural languages." Incommensurability involves meaning, but it ordinarily remains covert. For example, Einstein attached a new meaning to "simultaneity," which eventually enabled him to unearth some features of Newtonian physics that at a level of tacit awareness had influenced classical arguments of space and time. Since incommensurability depends on such covert premises and prejudgments that underlie broad conceptual frameworks, an explicit definition of those premises and prejudgments is hardly possible. The linguistic embodiment of a particular scientific cosmology consequently becomes, in the minds of its adherents, self-sufficient and self-confirmatory, and it entails a relatively elaborate belief system. This being the case, and contrary to Popper, even though refuting evidence presents itself, the scientist generally tends to persist in her belief, sometimes dogmatically, while conveniently sweeping the counterevidence under the rug. Or she might explain the counterevidence away by means of appendages she conveniently attaches to the received conceptual framework. And all the while, language (that is, the meanings of the terms she uses in describing that selfsame theoretical framework) continues to be

regarded, whether tacitly or consciously, as self-sufficient and self-confirmatory.

Suspended within Language

The upshot of all this is that, following the RMV thesis, change of scientific theories and of the language of science does not come about because science discovers new facts. Science adopts new linguistic formulations of the old facts, scientists become at home with them, and subsequently the facts are construed in an entirely different light (Feyerabend 1975, appendix 5). If scientific facts are so construed, then the linguistic formulation of the old facts in the previous theoretical framework becomes incommensurable with the linguistic formulation of the new facts. There is simply no way of translating between the two languages faithfully, nor of comparing and contrasting them from within their respective frameworks. "Mass" from within Newtonian physics has one particular meaning, which, from within, is construed to be "true," and if this Newtonian meaning is placed within the Einsteinian framework it is not as a consequence rendered "false." Rather, it becomes merely absurd, nonsensical, or virtually meaningless. To be considered "true," it must first take on a new meaning commensurate with the entirety of Einsteinian physics. Hence there are no available Newtonian grounds for accepting Einsteinian relativity, nor are there Einsteinian grounds for embracing Newtonian physics. In other words, the Einsteinian cannot declare that Newton was mistaken in the calculation of his time/velocity equation. The most he can say is that different equations were used, which form the foundations of the Newtonian universe. It is not a question of which discourse best describes "real world" phenomena; it is simply that the operations of each discourse produces a distinct body of "knowledge" exclusive to a particular scientific community. The operations should not be considered dependent either upon ontological or epistemological stipulations.

In this respect, a scientific "paradigm" in the manner of intertextuality is a seamless fabric, a particular thread of which, when pulled, either tears loose or drags the entire fabric along with it, or everything becomes completely unravelled, depending upon which thread is pulled, who does the pulling, and how it is pulled. And like intertextuality, language, both formal and natural, is the medium. The conceptual framework within a particular "paradigm" is generated by it; but there is no method for speaking *about* its grounds, its foundations, for a language about grounds can be none other than a groundless language. We blissfully swim along in our language medium, while discovery of the

Chapter Seven essence of the medium remains beyond us, though at least some of the rules for swimming do not. The same can be said of a particular conceptual framework. For its believers, trapped inside, it is the only viable one that can be envisaged, and under ordinary circumstances they cannot effectively get outside to see alternatives or to ask themselves whether other conceptual frameworks are better or worse. Hence from within a particular linguistic perspective or frame of reference, and following Heisenberg, the scientist cannot say language is *like this* or *like that,* and therefore he cannot say the world is *like this* or *like that.*[5]

However, if two conceptual frameworks or "paradigms" are presumably incommensurable, then one might propose construction of a more general language girding both and by means of which it becomes possible to adjudicate between them. There can be no such neutral metalanguage, says Feyerabend, the most outspoken of the RMV theorists. Nor is a better theory of meaning the answer. In fact, Feyerabend argues that we must throw out theories of meaning altogether. We should contemplate only sentences qua sentences, attending to what is said and even how it is said, not what is meant. This is somewhat reminiscent of the later Wittgenstein (1953, 46) for whom the more narrowly we examine our actual language, the wider the chasm between it and our ideal becomes, so we retreat "back to the rough ground." Heisenberg is even more poignant:

> In the *Tractatus,* which I thought too narrow, [Wittgenstein] always thought that words have a well-defined meaning. We can sometimes by axioms give a precise meaning to words, but still we never know how these precise words correspond to reality, whether they fit reality or not. . . . I would hope that philosophers and all scientists will learn from this change which has occurred in quantum theory. We have learned that language is a dangerous instrument to use, and this fact will certainly have its repercussions in other fields, but this is a very long process which will last through many decades I should say. (Buckley and Peat 1979, 11–12)

Tradition has usually held that behind and beyond language there is univocal meaning: here is the word and there is the meaning, which can be disclosed to the mind with proper training and insight. Now, it appears, whether we are speaking of scientific or any other discourse, nothing is to be expected above and beyond language, that is, words.

Interestingly enough, a notion akin to incommensurability, or quasi-incommensurability, exists in certain of Borges's

texts. Nonetheless, apparently some degree of communication across the supposedly bottomless chasm *does* occur, as does at least partial comprehension of an alien conceptual framework. In Chapter Three, it was pointed out that Averroes rather effectively, though incompletely, intuited what would ordinarily be conceived as a segment from a form of life incommensurable with his own: Greek tragedy and comedy. And Borges, on writing his story, apparently demonstrates some insight into Averroes's Islamic culture, though, perhaps wisely, he denies it. Let us return briefly to this story in an effort to elucidate Borges's instantiation of an alien conceptual framework, even a form of life, potentially penetrated.

Suspended within Language

During Averroes's visit to Faruch's home, Abulcasim, the traveler, was asked to relate some of his adventures. What could he tell? he asked them. They demanded marvels of him, which were perhaps "incommunicable" (and this choice of words is not unimportant). Finally, after some cajolery, he began relating his bizarre experience in a "house of painted wood where many people lived" (a Chinese theater). Abulcasim attempted to describe this strange happening as if he were observing the inhabitants of a Chinese household, but it was rather impossible. There was a single room with rows of cabinets or balconies on top of one another. Some people were eating, some drinking, some praying, singing or conversing, while still others were standing on an elevated terrace playing drums and lutes. During his stay in this "house" Abulcasim experienced many curious events: people were thrown in prison, though it had no walls; others traveled on horseback, but there were no horses to be seen; still others were killed, only to stand up once again. Abulcasim's audience was amused; Faruch even called the scene an act of madmen. But these were no madmen, Abulcasim tried to explain further, with considerable "use of his hands," for he was now clearly in what he had feared to be an "incommunicable" situation. A Chinese merchant told him that they were representing or performing a story, instead of telling it in the customary manner. Faruch asked if they spoke. Of course they did: they "spoke and sang and perorated" (*L*, 153). To this, Faruch indignantly retorted that no more than one person was necessary, for a single speaker could say anything, no matter how complicated.

Everyone approved of this dictum, and they went on to extoll the virtues of their own Arabic, the language "God uses to direct the angels," and Arabic poetry was equally praised. Abdalmalik, who considered himself a worthy critic, labeled the pastoral poets of Damascus and Cordova as antiquated; he

Chapter Seven   considered it absurd "for a man having the Guadalquivir before his eyes to exalt the water of a well" (*L,* 153). He continued, arguing that unique metaphors move their readers initially, but a few centuries of admiration render them valueless. Abdalmalik offered the example of the ancient poet, Zuhair, who in his *Mohalaca* compared destiny to a blind camel that suddenly tramples men into the dust. Such a figure could once move people, but no longer, for the metaphor had outworn its use. Averroes remained silent. But evidently certain cues during the conversation had brought about a fruitful conjunction. Earlier, Abulcasim had observed that all things were mentioned in the venerable Koran. Faruch had argued that rather than the strange Chinese house replete with collective palaver, a solitary person sufficed to say anything and everything.

Finally, Averroes spoke, "less to the others than to himself." He had once defended Abdalmalik's view, but no longer, for two reasons. First, if the purpose of a line of poetry is merely to surprise us by an appealing linguistic turn, its life span would not cover centuries, but days, hours, or perhaps merely the time of its reading. Second, a good poet is not an inventor but a discoverer. Time enriches rather than impoverishes verses because their discoverers reveal something that simply could not have been other than what it was. Averroes argued that, in this respect, the first poets had "already said all things in the infinite language of the deserts," and in the Koran "all poetry is contained." With respect to Zuhair, Averroes continued, relationships between objects in nature—fruit associated with green birds, for example—are relatively effortless. There are "infinite things on earth; any one of them may be likened to any other" (*L,* 154). This activity is arbitrary, a mere game. In contrast, the correspondence between words and the furniture of this world is much more difficult; understandably, writing should be considered as art. Averroes then introduced an abstract noun, "destiny," to illustrate his point: "there is no one who has not felt at some time that destiny is clumsy and powerful, that it is innocent and also inhuman" (ibid.). Nobody can elude this conviction, which Zuhair first incorporated into a poem, and the poem will never be superseded, for time "enriches verses." Averroes then argued that the poem now served to evoke their memory and relate the misfortunes destiny had dealt them with those of the deceased poet. Zuhair's figure including the old camel and destiny originally had two terms. Averroes proposed that it now had, in the context of their situation, four terms: Zuhair's des-

tiny, his camel trope, their own destiny, and their interpretation of the trope. In conclusion, Averroes condemned those illiterate and vain seekers of trivial innovations; rather than being true poets, they merely engaged in child's play. Everyone enthusiastically approved, for Averroes "was vindicating the traditional."

And the improbable had transpired. Though he knew not how, something had revealed to Averroes the meaning of the two obscure Greek words. After returning to his library, and with careful calligraphy, he equated *tragedy* with "panegyrics" and *comedy* with "satires" and "anathemas." The narrator, to repeat, then intervenes, revealing that he had attempted "to narrate the process of a defeat."

The narrator's dilemma, in light of the incommensurability thesis is twofold. In the first place, the problem of translation enters the picture. Averroes depended on an Arabic translation of Aristotle to interpret Greek literature, and Borges (the narrator) resorts to secondary sources—the works of Ernest Renan, Miguel Asín Palacios, and Edward William Lane—in his attempt to recreate Averroes's world. In the second place, the narrator confesses that his goal, like that of Averroes was monumental, if not impossible: in order to compose the story, he had to be Averroes, but in order to be Averroes, he had to compose the story, and so on. This regress corresponds to yet others. Earlier, Averroes had ruminated that for a person to become incapable of sin he must first have tasted sin. And, following Averroes's hypothesis of poetic discovery, the poet must create not the astounding, but what each person—which is to say, all persons—knows, but does not yet know he knows. This is, once again, the knowledge paradox.

It is also Platonic realism of the first order. In presenting Averroes's case, Borges by no means belies his occasionally professed nominalism, however. Like the reality of the Tlönians, when the narrator ceases to believe in Averroes, the Arabic philosopher, indeed, the whole of "Averroes' Search," no longer exists. Yet, the fact remains that Averroes somehow partly intuited a solution to his problem, and Borges, upon writing his story, somehow resolved his. In this sense, while engaged in their respective tasks, both men perhaps existed in some sort of imaginary Platonic form, but after the fact, they ceased to so exist according to the demands of nominalism. Borges in effect offers us a notable paradox *on* the learning paradox, thus revealing the logical impossibility of doing what we do naturally during the course of our everyday efforts to know.

Chapter Seven

All told, "Averroes' Search" is at once a narrative about (1) the problems of narrative, (2) the difficulty—though not the impossibility—of at least partly acquiring knowledge of the new, and (3) the equal difficulty of breaching the gap between incommensurables. Borges seems to suggest that a modicum of communication may be possible across apparently incommunicable chasms. One might justifiably argue that, though at best a translation across radically distinct—and perhaps even incommensurable—world-images and languages such as Spanish, Quechua, Arabic, and Japanese may be to a degree feasible, it is inevitably accompanied by a loss, for certain violence is done to the languages concerned. Nevertheless, we tend to persist in our effort to render a degree of communication feasible. Benjamin Lee Whorf, if correct, supposedly *was* able to convey key aspects of the Hopi language and world-image in English. Texts in European languages *are* daily translated into third-world languages, and vice versa, and they are traditionally "taught" in the classroom. A competent anthropologist presumably *is* able, with varying degrees of effectiveness, to understand the form of life and world-image of the culture she studies. And so on.

Feyerabend (1975, 223–85) argues that such communication across incommensurables does not entail comparability of meanings, but rather the capacity to move intermittently between mutually incommensurable languages and world-images.[6] For example, a native speaker of Spanish (Borges) and another of Arabic (Averroes) can to a degree communicate by Borges's interpreting Averroes's utterances in Spanish and Averroes Borges's utterances in Arabic. They will most likely muddle along swimmingly, each oblivious to the violence he may be inflicting on the other's language. Yet they are somehow communicating. In fact, two propositions, p (perhaps Aristotle's definition of tragedy and comedy) and q (say, a passage from the Koran) can potentially have some sort of *meaning approximation* toward a true *interpretation,* even though q is generally incompatible with Borges's world-image and p incompatible with Averroes's world-image (see Korner 1970, 63–65).[7] They discover they can *cooperate* with relative degrees of success.

The *meaning approximation* I refer to entails *information transfer* by way of the *core constituents* of a term or a set of terms that are partly common to two incommensurable conceptual frameworks, while *interpretation,* which cannot be absolute across incommensurable frameworks, entails *peripheral constituents* (i.e., incompatible terms, such as "simultaneity," from the

Newtonian to the Einsteinian framework). In other words, information from *core* concepts can be transferred between two incommensurables, which can then serve to *approximate* valid meanings in both, though interpretation, which includes *peripheral* and tacit meanings, cannot be made totally explicit across those same incommensurables. (This formulation falls in line with Polanyi's [1958] "tacit knowing" by means of *focal* [*core*] and *subsidiary* [*peripheral*] awareness, as I introduced it in note 10 of Chapter Two.)

For example, Averroes was correct insofar as both tragedy and panegyrics involve public assemblies and formal, serious, and laudatory discourse about a rather dignified individual who is generally conceived to be superior to the social norm. But he erred in that his analogy does not reveal the conflict between this person (who inherently possesses a "tragic flaw") and a superior force (his destiny), which leads ultimately to his downfall, thereby inducing pity and/or terror in the audience. Yet he was partly correct in this regard, since, when ruminating on Zuhair's trope he remarked that destiny, which seems to trample all people indiscriminately, brings to mind the readers' own misfortunes in such a manner that they can empathize with the dead Arab's plight. Averroes's coupling of comedy to satire and anathema partly hits the mark also, for both comedy and satire ridicule and deride human vices and follies (which are anathema, hence they are to be cursed and banned). But he fails to reveal the light and amusing nature of comic discourse, as well as its customary happy ending—though the Arabs' amusement concerning Abulcasim's experience and their ridiculing that bizarre Chinese practice was itself somewhat of a comic nature. Hence Averroes's writing in his manuscript that tragedies and comedies "abound in the pages of the Koran and in the *mohalacas* of the sanctuary" (*L,* 155) presumably corresponds to *core* terms and concepts which Averroes's world-image shares with that of the *Poetics,* though disjunctions prevail at their *periphery.*

However, in another sense, Averroes, in fact all those present in Faruch's home, missed the point altogether. They failed to recognize that tragedy and comedy are dramas, produced on stage; they were unable to grasp the role of actors and spectators. Of course for the Chinese, the very notion of a play is most tacit in their consciousness. It is simply *the way things are,* and hardly anybody would pay this immediate fact any mind. If Abulcasim's Chinese informant was not explicit in describing Chinese theater to him, it is for basically the same reason that the word "camel" is

Chapter Seven absent in the Koran: camels, being ubiquitous in everyday living, warrant no mention. Rather ironically, before his appointment with Faruch, Averroes had received various subtle clues suggesting the nature of drama. While in his study, and baffled over the pair of Aristotelian terms, he overheard some children who, in "the *vulgar* dialect, that is, in the incipient Spanish," were spontaneously acting out a Moslem prayer ritual, some assuming the part of actors and others of the audience. Steeped in his Islamic tradition, however, Averroes paid the children's game no mind and continued his search among the books in his library.

Averroes's oversight is indeed significant, in light of the *core* and *peripheral* (or *focal* and *subsidiary*) *constituents* I have briefly introduced, which also bear on the Aleph/Zahir pair. The Alephic vision, ideally a totalizing image, is forbidden to natural language and mere mortals, trapped within our three-dimensional world. Yet awareness of a lesser whole, a *Gestalt*, can be tacitly available (for example, a configuration of chess pieces for the master). In contrast, the focal (linear, sequential) Zahirian vision, bound within time and corresponding to the mnemonic function of the brain (i.e., this chess piece is now "here," now "here," etc.) does not lend itself to a one-to-one correspondence (interpretation) of the terms and definitions between the Islamic and Greek world-images. By means of the complementary and contemplative Alephic vision, however, there can be a spontaneous, unself-conscious, rather swimming form of implicit knowing that can be made only partly explicit.

This process of knowing is akin to an anthropological method, namely Clifford Geertz's *experience-near* and *experience-distant* concepts, which bears on Averroes's problem. An experience-near concept "is, roughly, one that someone—a patient, subject, in our case an informant—might himself naturally and effortlessly use to define what he or his fellows see, feel, think, imagine, and so on, and which he would readily understand when similarly applied by others." This is the Alephic vision. An experience-distant concept, in contrast, "is one that specialists of one sort or another—an analyst, an experimenter, an ethnographer, even a priest or an ideologist—employ to forward their scientific, philosophical, or practical aims" (Geertz 1983, 57). This is comparable to the linear Zahirian image. Geertz goes on to write that for one to grasp another person's experience-near concept, and to do so in such a way as to illuminate one's own experience-distant concepts, is a task "at least as delicate, if a bit less magical, as putting oneself into someone

else's skin." The task appears well-nigh insurmountable because experience-near concepts constitute a large part of one's tacit knowledge, and hence they are rarely recognized as concepts at all, much in the manner of Feyerabend's covert meanings in the language of a given theory (e.g., "drama," which is a tacit and rarely explicated given within the Chinese form of life). Precisely, it is the nature of experience-near

Suspended within Language

> that ideas and the realities they inform are naturally and indissolubly bound up together. What else could you call a hippopotamus? Of course the gods are powerful, why else would we fear them? The ethnographer does not, and, in my opinion, largely cannot, perceive what his informants perceive. What he perceives, and that uncertainly enough, is what they perceive "with"—or "by means of," or "through" . . . or whatever the word should be. (Geertz, 1983, 58)

Geertz reveals that in the field he has arrived at his most intimate of notions not by imagining himself to be someone else—a rice peasant, a tribal sheikh—and then thinking (in Zahirian fashion) about what the other person thought and felt. Rather, the path entailed "searching out and analyzing" symbolic forms—words, images, behavior, institutions, in short, a form of life—in terms of how people actually represent themselves to themselves and to one another. This would involve, properly, and rather without thinking *about* the process, an oscillation between experience near and experience distant, and between *core* and *peripheral constituents,* (or *focal* and *subsidiary* awareness).

In this respect, while the small group of interlocutors in "Averroes' Search" extolled the virtues of Arabic ways, Abdalmalik significantly referred to the mundane, the commonplace—the water of a well, a blind camel—upon labeling as antiquated the poets who adhered to pastoral images and a Bedouin vocabulary. These are ordinarily for the Arabs experience-near concepts, which are deemed relatively unworthy of linguistic embodiment, much like the concept of the theater for the Chinese. In contrast, Abdalmalik would applaud the poet who evokes experience-distant concepts, that is, unique images, and who, like Geertz's "analyst, experimenter, ethnographer and even priest or ideologist," employs these images to promote his craft. Averroes's refutation of Abdalmalik represents a defense precisely of such experience-near terms in his own form of life, which could have aided him in tacitly relating Chinese theater to his

enigmatic Greek terms, but they did not. Averroes embedded the two terms in the Koran, where "tragedies and comedies abound." He remained incapable of the proper oscillation between frameworks.

A relativist anthropologism, and perhaps even some RMV theorists, would rule out the above connections between terms, definitions, and frameworks by means of *core* and *peripheral constituents* alike, claiming that there can simply be no linguistic mapping between them. In contrast, I suggest, and Borges's story seems to illustrate, that an increase of *information transfer* by way of the *core constituents* can yield at least the most adequate, though always incomplete, *interpretation* under the circumstances. This, I believe, would at least be Borges's elegant wish, judging from the many attempts in his stories to make the unthinkable thinkable and the unsayable sayable.

Words and chess pieces are analogous; knowing how to use a word is like knowing how to move a chess piece. . . . I do not deny that there is a difference, but I want to say that knowing how a piece is to be used is not a particular state of mind which goes on while the game goes on.
—Ludwig Wittgenstein

He who understands a baboon would do more toward metaphysics than John Locke.
—Charles Darwin

The Rules of the Game

2 The attempt described in the previous section to bridge the gap between incommensurables obviously shares common ground with hermeneutics. In fact, both Polanyi and Geertz demonstrate that they are knowledgeable of and influenced by the hermeneutical tradition. And perhaps rather surprisingly, for Kuhn, science itself is a sort of hermeneutical enterprise. Kuhn tells of his initial difficulty when attempting to understand Aristotelian physics. At the outset he could not to his satisfaction arrive at a proper interpretation. He became deeply perplexed because Aristotle, who was usually an acute observer

and had provided penetrating interpretations in biology and political behavior, made what appeared to be utterly absurd statements about motion. The more Kuhn read, the more confused he became. Finally, he focused attention on an apparently absurd passage, asking himself how a grown and presumably mature man could have written it. During these mental labors, and eventually, by somewhat of a "hermeneutical leap," he saw the text anew such that all its central passages now took on different meanings. His world-image had somehow partly fused with that of ancient Greek physics (Kuhn 1977, xii). Lest my sketchy exposé be misleading, I must add that Kuhn's experience would have been virtually impossible for a person without a vast background knowledge and extensive interpretative practice (even with such knowledge and practice, the task might appear insurmountable, judging from Averroes's dilemma). On the other hand, I have not referred merely to an isolated incident. Feyerabend's "anthropological practice" regarding scientific activity is well known as is, following the later Wittgenstein, comparable theory and practice by philosophers and social scientists of the stature of, in addition to Geertz, Richard Bernstein, David Bloor, Charles Taylor, and Peter Winch.[8]

These rather common experiences introduce us to a much more problematic situation than that confronted by Averroes and Borges. I refer to the narrator of "The Immortal," briefly discussed in Chapter One, and his effort to understand the strange Troglydite. Averroes attempted translation between two "incommensurable" languages and forms of life. In contrast, the Troglydite's silence does not allow the narrator any verbal clues whatsoever until their final scene together. After the narrator's encounter with the Troglydite, he becomes intrigued with him. He observes this curious being for some time, eventually coming to know him intimately. He gives him the name Argos, after the dog in the *Odyssey*. Initially, Argos remains a complete mystery:

> I thought that Argos and I participated in different universes; I thought that our perceptions were the same, but that he combined them in another way and made other objects of them; I thought that perhaps there were no objects for him, only a vertiginous and continuous play of extremely brief impressions. I thought of a world without memory, without time; I considered the possibility of a language without nouns, a language of impersonal verbs or indeclinable epithets. (*L*, 112)

Suspended within Language

Chapter Seven

After prolonged speculation, the truth is finally revealed to the narrator: during a touching scene, the Troglydite, "as if he were discovering something lost and forgotten a long time ago, . . . stammered these words: 'Argos, Ulysses' dog' " (*L*, 113). The Troglydites *are* the immortals. That is why language is virtually lost to them, and why they have relinquished any and all attempts to progress, to better themselves and their environment. After judging that all undertakings are in vain, they resorted to living in their thoughts. Now, absorbed in pure contemplation, they hardly perceive the physical world, nor are their thoughts linear but rather merely a haphazard collection in the timeless present. Nothing can occur only once for them; everything must occur an infinity of times. The immortals, in short, are pure quietists—the narrator recalls an individual he never saw stand up, and a bird had nested on his breast. We are told that the immortals, like the animals, live timelessly, but unlike irrational beasts, they know they are immortal, which is at once "divine, terrible," and "incomprehensible."

Language is of little consequence in the situation confronting the narrator. Any communication across the radically distinct orbs he and Argos inhabit must be for the most part extralinguistic. The chasm is even greater than that between, say, the aged Borges and the young Borges of "The Other." Or between our world-image and Carlos Castañeda's when, after his apprenticeship with Don Juan, he might proclaim: "People can fly without the aid of machines." Or when the schizophrenic tells us his body is a piece of junk, that it is not part of him. Generation gaps, myths populated with soaring *Homo sapiens,* and disembodied spirits have at least some relevance to our world-image. The chasm between the narrator and Argos, in contrast, is more akin to a flatlander confronting a spherelander—or, recall Ouspensky's quote in Chapter Six revealing his attempt to capture the essence of the fourth dimension.

Wittgenstein might have come close to the narrator's problem when musing that if a lion could talk we would not be able to understand it. Or when, for example, he writes: "What would a society of all deaf men be like?" (1970, 371). Or perhaps, a society of blind humans. Indeed, in a world inhabited by blind people, many or most of our customary language games would not be played. Their mind-set would be so radically distinct from our own as to render communication well-nigh impossible, above all, since our form of life is firmly embedded in language. In other words, in a country of the blind, the one-eyed man

would most likely be neither a king or a participant, but merely a spectator. Wittgenstein (1953, 90) tells us elsewhere: "We feel as if we had to *penetrate* phenomena: our investigation, however, is not directed towards phenomena, but, as one might say, toward the '*possibilities*' of phenomena. We remind ourselves that is to say, of the *kind* of statement that we make about phenomena."

Suspended within Language

Wittgenstein implies here that the "existence" of certain phenomena depends upon their "possibility" contained within language. For instance, the language game of science treats Eddington's writing desk as a swarming confusion of largely vacuous molecules; our everyday language treats it in quite another manner. What will be considered "possible" and what not depends upon the grammar of a particular language game; it depends upon what the grammar permits. The quantum theoretical view of Eddington's desk is largely forbidden to everyday language games; that is, its "possibility" of description in a radically distinct language is minute, yet a certain "possibility" exists, depending on the changes eventually forged in everyday language.

As a case in point, by means of "The Immortal," a reader can contemplate the characteristics of what might be the perspective of a being existing in a world radically distinct from her own. One might suppose, however, that this reader will be totally incapable of understanding it, for the chasm between the immortal's world and hers is simply not reducible to the "possibilities" contained within her language. Nevertheless, a certain approach, a static and rather languageless form of existence, might be to a degree possible, for in our contemporary society there exists a vague counterpart to the Troglydites' world. I speak of the avant-garde art briefly mentioned above. This art in general creates an aura of timelessness: purpose and expectations no longer obtain, there is neither good nor evil, redemption nor sin, joy nor sadness, hate nor love. A pure form of quiescent existence prevails, without dichotomies or hierarchies. All is flattened to a static present. The *ideal* becomes a sort of *satori*, when the oneness of the universe is potentially revealed. Leonard Meyer (1967, 169) elucidates a paradox in this philosophy of the avant-garde artists. They strive for an *ideal* expression without goals, reference, or meaning. But the very rules of the game prohibit such an *ideal*. The project is finished before it begins, for there can be neither beginning nor end. The same paradox strikingly inheres in the impossibility of

Chapter Seven the narrator's comprehending the mute Immortal's sphere of existence. Yet, intuitively linking one's wavelength to that of "unsayable" avant-garde art is somewhat comparable to the task of comprehending the world of the Immortal. The project is at least partially realized, though it be logically impossible.

Much like the Troglydites' world, neither is the quantum world nor the four-dimensional space-time continuum directly translatable into our own world-image as it is expressed by classical logic and natural language. Yet, it is not amiss to suppose that the possibility exists, by an imaginative leap, of at least approaching an understanding of these "incommensurables." Consider, for example, the timeless, static Minkowski "block" outside any and all three-dimensional observers, which does not happen, it simply *is*. Recall from the second section of Chapter Four that mathematicians, using computer mock-ups, claim they can get a certain "feel" for the fourth dimension. This is perhaps tantamount to the "feeling" lying behind and beyond some of Einstein's thought experiments, or the flatlander intuitively grasping the essence of Sphereland.

In this vein, what might be the perspective of a hypothetical four-dimensional observer from outside the "block"? Her perceptual grasp being all-inclusive and nonpositional, she would have no a priori notions either of time or of space. She would be James Jeans's ultimate Mathematician, capable of seeing anything describable in mathematical symbols as if in a timeless instant: nothing would rest outside her purview. How, for instance, would she perceive our game of chess? It would be for her a set of moves in three-dimensional space, but she would not "read" a single game linearly, through time; it would be a massive superposition of all possible moves and hence of all possible games in simultaneity. Whether considered a chronology of successive configurations on one board or a matrix of states on an exceedingly large number of boards stacked one upon the other would make no difference to her. In other words, we can see events successively in three-dimensional space, but the fourth dimension, time, is comparable to a thin slit affording us no more than successive slices out of the chess game as it proceeds. For our omniscient Mathematician, on the other hand, the time dimension from the beginning to the end of the game would be copresent, as would be our gaze of a flatlander's world. Nonetheless, though we cannot transcend our time-bound existence, our acquiring a vague "feel" for her universe by way of an analogy is not out of the question.

Suspended within Language

Skepticism might prevail, however. In the case both of "The Immortal" and the "block," translation (transportation) directly from one world-image to another must be enacted irrespective of natural language, for language is incapable of describing one of the two world-images. And understanding of a world-image unstructured by linguistic mortar is impossible, for we are *always already* "thrown" into a language and world-image. Whether we like it or not, we are profoundly influenced by our tradition. It stands over and against us: a set of prejudgments we tacitly acknowledge as *the way things are*. And it forges our language and our conceptual framework governing everything that can be seen and said from a particular vantage point. However, world-images, handed down to us by tradition, are finite and restrictive, yet they are incessantly changing. We cannot merely jump from our world-image to another incommensurable image; we are grounded in our own cultural situation. But we can to a degree "oscillate" between *core* meanings and *peripheral* meanings, whereby we gain some insight into an alien world-image and at the same time we enrich our own, during the process potentially altering some of our fundamental prejudgments. At the heart of the incommensurability thesis there exists a historical openness, rather than timeless closure, of language and understanding. And communication across radically distinct conceptual frameworks and forms of life is partly possible, since we are constantly challenged, and we challenge ourselves, to understand what appears at the outset unintelligible. In so doing, we place the tacit givens of our own form of life in jeopardy, but the possible gains of greater understanding of both forms of life far exceeds any loss that might supervene.

One problem is, to repeat, that interpretation of an alien world-image can never be completed; the task is infinite. There can be no more than successive approximations toward an absolute interpretation. And since we have no way of knowing where we are along the infinite line with an infinity of ramifications, we cannot know with certainty when we are progressing and when retrogressing. Another problem is that the medium of all world-images is distinctively linguistic. We are caught within the language we use (or rather, that uses us), and hence our language, which, like the Library, is open to an infinity or quasi-infinity of possibilities, bars us from knowing a priori where its limits lie. So, to reevoke Borges, we are always inexorably left with the vulnerable word against the world.

Chapter Seven   *[W]e no longer have that primary, that absolutely initial, word upon which the infinite movement of discourse was founded and by which it was limited.*
   —Michel Foucault

Texts of Our Own Making

3 Indeed, can the world continue to resist the word indefinitely? Let us return to the Library of Babel. Even though a particular symbol from the collection might be graphically identical to itself in all its uses, a potential infinity of possibilities is open to it, given a potentially unlimited number of textual contexts and contexts of readings. A given collection of symbols combines into words (whether meaningful or nonsensical), words into sentences, sentences into paragraphs, paragraphs into texts. Comparably, the universe—which is another name for the Library—consists of the combination of a few simples, which make up larger and more complex particles. These are the symbols of the "language" of nature, which are strung together with their own grammar rules: quantum theory. And they eventually form sentences (molecules):

> Soon we have books, entire libraries, made out of molecular "sentences." The universe is like a library in which the words are atoms. Just look at what has been written with these hundred words! Our own bodies are books in that library, specified by the organization of molecules. The universe as literature is, of course, a metaphor—but the universe and literature are organizations of identical, interchangeable objects; they are information systems. (Pagels 1983, 255)

The ancient topos of the universe as a "book"—or better, as a "text"—became an obsession with the Kabbalists and the alchemist Paracelsus,[9] whose "book of nature" could presumably be deciphered if read in the proper manner. The concept was revived in science, especially with Galileo (1957, 237–38), who declared that

> this grand book, the universe . . . stands continually open to our gaze. But the book cannot be understood unless one first learns to comprehend the language and read the letters in which it is composed. It is written in the language of mathematics and its characters are triangles, circles, and other geometrical figures without

which it is humanly impossible to understand a single word of it, without these, one wanders about in a dark labyrinth.¹⁰

Notably, Galileo envisages the book of nature to be written in mathematical language. However, whether in a formal language that inextricably loses its precision when interjected into the sphere of human intuitive faculties, or in a figurative or natural language irreducible to formal conciseness, the book of nature, as Heisenberg remarks concerning atomic physics, has taken on the trappings of an archetype (Miller 1986, 259).

Borges remained fascinated with the book of nature image, to which Mallarmé also alluded in his pronouncement that the world exists so it can be put in a book. On various occasions, Borges refers to Bacon, Bloy, Carlyle, Novalis, and de Quincey, all of whom registered more than a passive interest in the book metaphor (*OI*, 119; 125). He notes the Christian idea that the divinity wrote a book, or rather, two books, one of which was the universe (*OI*, 119). In "The Secret Miracle" and especially "The God's Script," God is a single letter, and the world is that letter generated into a single book, which contains God as an immanent being in the Spinozistic sense. The protagonist of "The Garden" speculates on the possibility of an infinite book and an infinite maze, both of which are one, and one with the universe. Uqbar is described in only one encyclopedia, in which there is bookish reference to the imaginary Tlön, which later takes on a "reality" eventually to become the world: a text. And "Undr" (*BS*, 81–87) and "The Mirror and the Mask" (*BS*, 75–79) entail a search for the ultimate expression of all things with the power of a solitary word.¹¹ The examples continue. Suffice it to recapitulate the views of two Borgesian critics: the predominant theme of Borges's fictions is that literature is as much being as the world (Lefebve 1964) and, if the world is a book, then all books are the world (Blanchot 1959).

Now the problem is that if the unruly interrelated jungle of texts is tantamount to the world, the world we are in, and if we are "always already in the text," wherever we are, then uneasy consequences undoubtedly follow for those who persist in nurturing nostalgia for the Golden Age of certainty. Given the problematics of language, if the world is a text, then the same problematics that plague language inhere in the world. For Borges, a "bookish" writer if there ever was one, this is nothing new. He approaches books before "reality." Dream precedes the world, a conceptual gaucho stands between him and actual

Chapter Seven   gauchos, imaginary tigers between him and actual tigers. Language is the mediator and moderator by means of which Borges longs for "reality," but this aspiration is self-defeating, and he knows it, for the power of the word is a long-lost hope. Can the word and the thing, acts of mind and concrete experience, be reconciled? That game is illusory altogether for Borges. But purely mental constructs can become *the reality* appropriate for fictions, since they imply a realism of the only knowable kind, the *reality* of the mind.

Borges would concur, then, with Paul de Man (1971, 17) that literature is always mediated, always a step away from the world. There is no necessary reference language, hence "truth" is so elusive that we can do no more than settle for fragmentary visions of the world. Borges is not as fatalistic regarding mental activity as he is about physical "reality," yet both are ultimately deficient. A constant theme in *Dreamtigers* is language's deficiency and the hopeless dream of its mirroring the world. Words can do no more than mirror themselves, and, at most, they mirror us. Like Eddington's footprint in the sand, and whether speaking of the scientist or sage, fabulist or philosopher, the impending realization is imminent: the world, our world, is an image of ourselves. Unfortunately, however, our thought often remains muddled due to our taking language as a model or picture of the way the world is.

Although strings of verbal or written signs can appear to have reference, this is not necessarily so; or perhaps we should reply, this is hardly ever so, if at all. Support for such a reply abounds in Borges's texts. For example, after playing the role of a prostitute, Emma Zunz existed in a "time outside of time," in an incredible story, "but it impressed everyone because substantially it was true. . . . [O]nly the circumstances were false, the time, and one or two proper names" (*L*, 137). The only witness to the killing is dead, the sailor is now divorced from Emma Zunz's world, and the account remains divorced from "reality," its meaning depending not on the physical world but on an imaginary perspectival framework. The strings of words in "Emma Zunz" have the quality of "as if" hypostats, but such hypostatical acts threaten to become a falsification of language: words beaten into a different shape in order that they signify what they would otherwise not signify; or perhaps, in order that they signify what could otherwise have been signified in an indeterminable number of ways.

Regarding this language-dependency of "reality"—the core of the RMV theory—reconsider for a moment "Pierre

Menard." The two orthographically identical texts of Cervantes and the French writer are, according to their readers, nonetheless diametrically opposed to one another, much like two incommensurable scientific theories. Cervantes's praise of history is conceived to be mere rhetoric; Menard, in contrast, "a contemporary of William James, does not define history as an inquiry into reality but as its origin. Historical truth, for him, is not what has happened, it is what we judge to have happened" (*L*, 43). Cervantes's style is "the correct Spanish of his time"; Menard's is "quite foreign," "archaic," and "suffers from certain affectation" (*L*, 43). Cervantes's text, "in a clumsy fashion, opposes to the fictions of chivalry the tawdry provincial reality of his country" (*L*, 42). Menard, in contrast, "selects as his 'reality' the land of Carmen during the centenary of Lepanto and Lope de Vega" (*L*, 42). In short, a reading of each of the two texts is a reading of two worlds. This represents a sort of "naive textual realism." Identical texts in distinct contexts are conceived to be radically distinct. That is, iteration (rereading) of a text becomes, rather than a difference, *pace* Derrida, something partly to wholly incommensurable.

Suspended within Language

Moreover, if the same text by two different authors can become two contradictory texts, then one might suppose that the historical context of those texts' writings and readings becomes all-important. At any given moment in time, writers and readers are restricted by their finite limitations, but over the broad expanse of time, nothing is impossible. Yet, as Menard once wrote to the narrator: "Every man should be capable of all ideas and I understand that in the future this will be the case" (*L*, 44). Hence there is, in effect, no constraint preventing one from interpreting Menard's text as if it were Cervantes's or vice versa. Or from interpreting, in light of Borges's style of intertextuality, Kafka as Cervantes, as Zeno, as Kierkegaard, as Browning, as Menard, or even as Borges. Or from interpreting the same "text" of nature differently (e.g., Priestly persisting in calling oxygen "dephlogisticated air," while Lavoisier termed it more properly). Or from reading a scientific text as fiction (which it inevitably is, since, if not already falsified, it will be in the future, and if not, then it was not scientific in the first place—at least according to Popper).

Very roughly, such "schizophrenic" readings of texts, or of the "text" of the world, are patterned on a variation of Wittgenstein's use of the "rabbit-duck" icon. The same set of marks can be labeled either one or the other, according to the disposition of their reader. On the other hand, if the drawing is placed in the

Chapter Seven

context of rabbitlike characteristics, the reader ordinarily disposed to see it either way will almost invariably report it to be a "rabbit." And vice versa. It was pointed out above that the matrix "representation" and wave "representation" of the particle-wave are incompatible, but at a deeper level they form part of the same unifying perspective. Regarding Cervantes's and Menard's texts or Wittgenstein's "rabbit-duck" example, the opposite inheres: though identical phenomena exist on the surface, there is a mind-generated underlying incompatibility. Incessantly altering contexts breed invariably differentiating readings and therefore differentiated worlds.

According to the RMV theorists, Newtonian and Einsteinian physics are inscribed as two distinct texts, and the conjunction of both generates an incommensurability. This situation might appear to be totally distinct from that of Cervantes and Menard. "History" as Cervantes conceives the term is incommensurate with Menard's concept of "history." However, just as terms like "mass," "space," "time," and "energy," now from within the Newtonian framework, now from within the Einsteinian framework, take on incommensurable meanings, so Menard's meaning of "history," when attached to the term wherever it occurs in Cervantes's text, becomes not simply erroneous or contradictory but absurd, nonsensical, or even meaningless. And the same inheres when placing Cervantes's meaning of the term within the context of Menard's text.

Borges, no less than Feyerabend and the RMV theorists in general, warns us about the limitation of the word and the poverty of our conception of the world. The Library and other Borgesian constructs demolish the Newtonian universe which gives rise to various antinomies. But Borges does not want yet another "reality" merely to replace it. Rather, he juxtaposes a new loss of certainty with old certainties to render everything uncertain. Hayles (1984, 151) remarks that in general, Borges's fiction "differs from scientific models of the field concept in using the concept that everything, including reality, is a fiction," while scientific models, in contrast, "are useful only because they are presumed in some way to reflect reality." Though her essay is in my estimation brilliant, Hayles is, I believe, slightly mistaken here. Nietzsche, Vaihinger, and the RMV theorists, as well as most of the "new physicists," properly view all models as fictions. Or, to paraphrase Heisenberg (1958a, 58), our theories are not of "reality" but of the interaction of our mind with the universe.

Rodríguez Monegal (1978, 138) and Barrenechea (1965, 79–83) cite Borges's debt to Fritz Mauthner.[12] Whether or not Borges was influenced by Mauthner is irrelevant. The important issue is whether in the German thinker there exists another possible tie between the Argentine writer and the RMV theorists, for Mauthner, like Feyerabend and others, believed language to be a self-contained whole. According to Mauthner, we always move within the limits set for us by the prevailing linguistic conventions. Even when we attempt to argue against the current view, we must use the presuppositions of that view, for there is no absolute escape from conventional practices. We might possibly be transferred from one prison-house to another, but we cannot totally free ourselves; we cannot get out of language by the use of language.

An interesting aspect of Mauthner's critique is that language entails three conflicting perspectives, three pictures of the world: the *substantival,* the *adjectival,* and the *verbal* (Weiler 1970, 282–88). The first of these pictures in the order of experience is adjectival. This is the world we know directly through the senses, and beyond that, our perceptions and value judgments (i.e., what we call beautiful, good, correct, etc.). This picture is not yet an interconnected whole; it is a mere aggregate of isolated experiences. It is the immediate raw material of the world only minimally conceptualized. Once these aggregated adjectival entities are interrelated in time, the verbal world becomes possible. This world does not believe in the substantival orb and it is not satisfied with the adjectival; it "sees in everything only change (for which alone it cares), relations, relations of the so-called things to us and relations of these things to one another" (quoted in Weiler 1970, 283). The picture of verbs, a world in incessant movement, is the subject matter of the physical sciences and cannot be complete in itself. The adjectival picture gives rise to the question, What is it that adjectives qualify?, and the verbal question is, What are the relations relations of? The third picture, the substantival, now becomes necessary. This is the realm of being, the condition of which is space as a complement to the temporality of the verbal picture. Mauthner believed that the differences between art, science, and mysticism correspond to the differences between the adjectival, the verbal, and the substantival. He even speculated what it would be like to construct three different languages, one for each world picture, concluding that these languages would be incompatible, if not to say thoroughly incommensurable.

Chapter Seven

Of course, we are already familiar with this story from Borges's "Tlön," where he encapsulates his linguistic preoccupations in the adjectival language of the Northern Hemisphere and the verbal language of the Southern Hemisphere, a Platonic form of substantival language being generally ignored. The first language affords a nonlinear set of static images. To cite Borges's example, "moon" would be "round airy-light on dark" or "pale-orange-of-the sky." And the second language would consist of a stream of ceaselessly fluctuating images; no word is the equivalent of "moon," but only verbal correspondence, which in English would be "to moon" or "to moonate." Hence, "the moon rose above the river" would be "upward behind the onstreaming it mooned" (*L,* 8–9). Borges also refers to another language in the Northern Hemisphere, the only form of substantival language possible for Tlön's inveterate idealists. This language, existing in their literature, which, it will be recalled, is the equivalent of their metaphysics

> abounds in ideal objects, which are convoked and dissolved in a moment, according to poetic needs. At times they are determined by mere simultaneity. There are objects composed of two terms, one of visual and another of auditory character: the color of the rising sun and the faraway cry of a bird. There are objects of many terms: the sun and the water on a swimmer's chest, the vague tremulous rose color we see with our eyes closed, the sensation of being carried along by a river and also by sleep. These second-degree objects can be combined with others; through the use of certain abbreviations, the process is practically infinite. . . . The fact that no one believes in the reality of nouns paradoxically causes their number to be unending. (*L,* 9)

The two languages of Tlön, like Mauthner's three perspectives within language, in a very real way suggests Borges's language dilemma, or that of any other competent writer for that matter. The ultimate nature of the world "out there" and of the writer's fictive worlds become one: Mauther outlines our entrapment within language; the physicist in the final analysis sees his own footprint in the sand; the writer, after years of effort, discovers that the labyrinth of lines he has patiently constructed "traces the image of his face." Both the physicist's "world" and the writer's "fictions" struggle to transcend the dualities of language. But to no avail, for language always wins the final round (see also Bunge 1987).

The multiple images (that is, metaphors) I have evoked in this inquiry force the conclusion of an unlimited iterability (read-

ability) of texts, of the universe (as "text"), or of the books housed in the Library (as the universe).[13] This conclusion is corroborated by the Borges-inspired "thought experiments" I have developed in the present chapter: (1) the partial decipherability of "incommensurable" texts and forms of life; (2) a languageless "text" that suddenly becomes profused with "textuality"; (3) identical texts that, when contextualized, become virtually incommensurable in the minds of their readers; and (4) the universe "read" by the Ptolemaics and the Copernicans, the Newtonians and the Einsteinians, ad infinitum, as an indefinite number of discordant "texts." On the one hand, if the same text can become many texts, whether we are speaking of the Menard-Cervantes text or the "text" of the universe for scientific inquiry, then many texts can become an indefinite number, if not an infinity, of texts. On the other hand, if incommensurable texts are at least partly intertranslatable, then there are not many texts but one text consisting of a plethora of mutually interpenetrating texts.

Suspended within Language

Perhaps that host of agonized readers scurrying about in the Library should take heart, giving in, finally, to the "Great Text" within which they are contained, and merely play the game of language, which in turn plays/will play/has played them. Borges implies so much.

My internal and external life depend
so much on the work of others *that*
I must make an extreme effort to
give as much as I receive.
        —*Albert Einstein*

4 It has become obvious that, in general, the present century has proved to be by and large an intellectually exhaustive period. The mathematician G. H. Hardy (1967, 85–86) once wrote: "I have never done anything "useful." No discovery of mine has made or is likely to make, directly or indirectly, for good or for ill, the least difference to the amenity of the world. I have helped to train other mathematicians, but mathematicians of the same kind as myself, and their work has been, . . . as useless as my own."

The Receding Horizon

Hardy's extreme statement reveals that the highest achievement of the mathematician is to produce a lasting work of art. If it proves to have practical applications, so much the better. Notice the similarity to Borges, whose aspiration is

Chapter Seven merely to provoke the reader with unexpected aesthetic juxtapositions. The quest is not for truth but for the untruths concealed in all presumed truths: to pull the thread that unravels the entire fabric, in short, to "deconstruct." Borges encourages us to turn the supposed river of knowledge back on itself, to think forward, backward, upside down, inside out, to see things intermittently from many perspectives in order to expand our mental outlook, which inextricably tends toward increased impoverishment: cerebral entropy.

This is a most proper activity. Following Mauthner and paralleling a host of contemporary critical theorists, to critique language, and to critique our own critique, is genuine critique. The problem is that our very exercise of such critique is itself a linguistic activity. Ultimately, there is no center, no fulcrum point by means of which a critical language can displace intertextuality. If true critique there be, it must lie outside language, but if so, then it cannot be accessible. In the Nietzschean sense, one might retort, the response should be laughter, the higher form of knowledge. Laughter is the spontaneous outburst when one is faced with an unexpected twist. Borges, of course, continually presents the reader with a surprise, with subtle, ironic humor. The commonplace becomes uncanny, the mundane miraculous, trivia becomes important, and vice versa. The reader remains for a moment with a puzzled brow, then a twitch of a smile shows, the head nods in agreement, the eyes twinkle. But an irremediable vestige of the initial perplexity remains.

Borges, it has become evident, wisely eschews meaning in the ordinary sense. There is no one and only hidden meaning behind a set of his sentences to be disclosed by the annointed priests and priestesses of interpretation. With respect to Borges, and many other contemporary writers, the very notion of a theory of meaning or interpretation should be abandoned; attention should rest on what is said, on sentences qua sentences, on the ongoing flow of the signifying process. This entails aspects of language that potentially lay bare the contradictions, ambiguities, paradoxes, and vagueness in its very usage, but there is nothing over and above what these devices are, nor can it be determined if they are valid or not: they simply *are*.

Borges does not ask the question, "Is what Hume says true and what Descartes says false?" He asks, "How do Hume's and Berkeley's dictums combine to refute time in such a manner that we persist in disbelieving this refutation?" Or, "How is dream like reality such that both are symmetrical, and why do we insist

on rendering them asymmetrical?" Or even, "What are the implications of finitude interjected into infinity, a series into a continuum, a text into the class of all texts, a man (or tiger) into the class of all men (or tigers) such that they are perceived both as equal and different?" Questions of this nature have nothing necessarily to do with determinate meanings, or with circular ruins, writers condemned to the firing squad, cerebral mnemonists, international spies, anguish-ridden library rats, and so on. Admittedly, texts cannot be categorically divorced from the world. If Borges were to write a story with absolutely no reference to what are conceived to be our "real world" objects, acts, and events, it could be none other than totally unlikely, unintelligible. Texts presuppose writers from whose pens they flow, the occasions of their writing, contexts of readings, institutions that sanction or condemn them, conventions that govern their future. But the interactions between writers, readers, institutions, and societies at large are not determined by meanings necessarily embodied in texts. Meanings are invariably stamped on texts, at particular moments, in particular situations.

Ultimately, for Borges it might appear that literature has no commendable future (Updike 1965, 12). Whatever could be said has been said; literature, civilization, the world, are exhausted.[14] Borges's remark regarding *Ficciones* is relevant: "I wonder if there is a single original line in the book. I suppose a source can be found for every line I've written, or perhaps that's what we call inventing—mixing up memories. I don't think we're capable of creation in the way God created the world" (Dembo 1970, 323).

This theme is very effectively reiterated in "Utopia of a Tired Man" (*BS,* 89–96). The narrator, while traveling in the pampa, is overtaken by rain and finds refuge in an isolated house occupied by a frighteningly tall individual. It is soon discovered that somehow the narrator has traveled thousands of years into the future. The two men's common language apparently being Latin, the narrator begins an inhibited conversation with his host. He learns that the earth's future inhabitants are taught in the schools to doubt—the art of forgetting—to shun pointless details. Though they are forced to live in time, which is successive, they never abandon their attempts to exist *sub specie aeternitatis.* The host, who is four centuries of age, has read no more than a half-dozen books, for, "rereading, not reading, is what counts. Printing—which is now abolished, since it tended to multiply unnecessary texts to the point of dizziness—was one

Chapter Seven of man's worst evils" (*BS*, 92). The language of this future race has become stark; quotations are all they have, nothing new remains to be said. In this world, like that of the Immortals, there is little or no hope of progress. Each man, at the age of one hundred, when he matures, "comes face to face with himself and his loneliness" and fathers one child, which is all that is necessary, since there "is no reason to carry on the human race" (*BS*, 93). At this tender age, one begins striving to forget the past, to reject anticipations, to live only in the here and now. This is indeed a spirit of exhaustion with a vengeance.

And yet . . . though the project is impossible, the quest intractable, Borges himself is never entirely free of his longings. He does not, he cannot, totally liberate himself from Spengler's archetype of "Faustian man," that singularly Western rugged individual, that laissez-faire entrepreneur, that prime motivator, who views himself as being trapped in a world of interminable obstacles and who incessantly reaches for the ultimate. We find the apotheosis of "Faustian man" in one who has so inspired Borges: Schopenhauer, along with certain of his contemporaries, Goethe, Beethoven, and others. The ultimate extrapolation of "Faustian man," however, might be found in Don Quixote; his counterpart rests in that representative of other quite mundane goals: Sancho Panza. The strain of upholding impossibly high ideals inevitably ends in anguish, frustration, hopelessness, quietism, or perhaps dementia. Borges, it seems, and, in spite of himself, would like to realize Faustian ideals (Platonism, mysticism, creation of the unique, determinate knowledge). He is therefore heroically condemned to the struggle, but with full awareness that there is no end point, nor can there be a turning back.

At my age (I was born in 1899), I cannot promise—I cannot even promise myself—more than these few variations on favorite themes. As everyone knows, this is the classic recourse of irreparable monotony.
—Jorge Luis Borges

Notes

1 Borges remarked in an interview with Jean de Milleret (1967, 116): "You want to make me into a philosopher and thinker; but the fact is that I repudiate all systematic thought because it always tends to deceive one."

Preface

2 For some general studies on parallels between mathematics, science, literature, and the arts, see Scott Milross Buchanan (1962), Martin Green (1963), John Gribbin (1979), Arthur Koestler (1964), Leonard B. Meyer (1967), I. A. Richards (1926), Wylie Sypher (1962), and in his own way, Douglas Hofstadter (1979).

3 In particular see Fritjof Capra (1975), P. C. W. Davies (1981, 1983, and 1984), Michael Talbot (1980), and Gary Zukav (1979).

4 Though I have some background in the sciences—I taught chemistry and physics in secondary schools for six years—perhaps a professional mathematician or scientist should have written this book. I, with some reservations, decided to undertake the task, asking patience regarding my shortcomings from the specialists in the diverse fields into which I venture.

5 I do not wish to leave the reader with the idea that I believe Borges to be everything to all people, or that anything and everything

Notes to Pages: xvi–7

can be read into his work. Certain other authors, such as Samuel Beckett, John Barth, Thomas Pynchon, and perhaps Robert Musil, could, I suggest, be placed under a comparable spotlight.

6   I have been selective regarding the scientific theories and broad speculations I have placed beside Borges's texts. For example, I have not addressed myself to the recent euphoria in certain circles over "superstrings" as a hopeful "Theory of Everything" (TOE) (see Davics and Brown 1988), to Ilya Prigogine's (1980; Prigogine and Stengers 1984) "order out of chaos" by way of "dissipative structures," or to the fascinating new "physics of chaos" (Gleick 1987; Briggs and Peat 1989). The latter in particular would appear relevant to Borges's allusions to randomness and disorder, especially in "The Lottery of Babylon" (*L*, 30–35) and "The Library of Babel" (*L*, 51–58) (It bears mentioning, however, that N. Katherine Hayles [1990] has recently published an excellent study relating "chaos theory" to works by Henry Adams, Stanislav Lem, Doris Lessing, and others). I have also directed no attention to the hopeful search, intensified in recent years, for a "Grand Unified Theory" (GUT), of which Stephen Hawking (1988) is one of the most prominent theoreticians. I do not offer these admissions as an apology, though certain regrets linger. Rather, given the overwhelming magnitude and complexity of the contemporary scene in the philosophy of mathematics and science and in theoretical physics, I have had no recourse but to pick and choose.

Chapter One

1   See Merrell (1980 and 1983) for a theoretical exposition of this concept.

2   This Plato/Aristotle distinction is Borges's, not my own. While it is to a degree valid, it is also strained. Actually, both Platonism and Aristotelianism share a fundamental premise: knowledge is final, impersonal, and certain. Platonic certainty is predicated upon eternal transcendental objects, Aristotelian certainty upon the invariant nature of things in the world. And both, as I illustrate below, have been superseded by the postclassical twentieth-century view.

3   However, in an interview with Jean de Milleret (1967, 157), Borges denied ever having read Vaihinger.

4   See especially Bruner (1957), Cassirer (1955–57), Gombrich (1960), and Piaget (1971).

5   Another interesting example of Borges's use of the coin symbol is his poem "The Iron Coin" (*OP,* 507).

6 Borges, in the journal *Sur* (1939, 13–16) speaks of Lasswitz's Library, which appears in one of his fantastic tales, translated and reprinted in *Fadiman* (1958, 237–47). And Abraham A. Fraenkel (1953, 6) refers to a work by E. E. Kummers on a comparable "Universal Library" that antedates that of Lasswitz. It is obvious that Borges has elaborated on a concept that has fascinated certain intellectuals over the past decades.

7 Gregory Bateson (1972, 177–93) believes the rudiments of this process to be found in certain animal communication, particularly evident in canine, porpoise, and otter play. See Merrell (1983) for further discussion regarding the function of negation in fictive constructs.

8 See Hayles (1984, 144–47) for a cogent discussion of this aspect of Tlön. And for a more detailed account of Wittgenstein's social interpretation of mathematics, see Phillips (1977, 119–68).

9 This is very roughly the general thesis of what has been called the "radical meaning variance" theory in philosophy of science held by, among others, Paul Feyerabend, Norwood Hanson, Thomas Kuhn, and to an extent Michael Polanyi, which I shall briefly introduce in Chapter Seven.

10 See, in general, Castañeda (1979), Chisholm (1973 and 1982), Findlay (1963), Parsons (1974 and 1980), Routley (1979), and Schultz (1979).

11 In this sense, nonexistent possibilities are rooted in our very language-using capacity. This notion, given the above assumptions, must embrace possible as well as actual statements in natural and artificial languages, and once again the total range of nonexistent sayable possibilities becomes limitless. Rescher (1975, 212), in a similar vein, tells us that "since there is nothing inherently finitistic about our linguistic resources, our thesis that possibilia inhere in the imaginatively projective proliferation of linguistic combinations does not impose upon them any conditions of finitude (or denumerability)."

12 Another story, "The Congress" (*BS*, 27–49), is comparable to "Tlön" insofar as a group sets for itself the project of mentally constructing a world. In Borges's own words, the story's subject "is that of an enterprise so vast that in the end it becomes confused with the world itself and with the sum of daily life" (*A*, 114).

13 This we shall observe further in the first section of Chapter Five from an article by Kurt Gödel on Einstein.

Notes to Pages: 30–49

14 See Bunge (1987) for a charming essay on the interface between fantasy in the arts and in the sciences, with special emphasis on Borges and Einstein. I would take issue with Bunge, however—as will become evident below—in his assertion that "scientific fantasy is justified solely insofar as it helps the scientist to comprehend the world" (Bunge 1987, 62; translation mine). Bunge, I believe, does not take sufficient account of the purely aesthetic dimension of theoretical science, as witnessed by works such as those of Arieti (1976), Augros and Stanciu (1984), Hadamard (1945), Miller (1986), and Wechsler (1978).

Chapter Two

1 It bears mentioning that Lorich does attribute excellence to Borges's "The South" (*F,* 167–74), since, like Julio Cortázar's short story, "Axolotl," it does not merely "depend upon convoluted abstractions."

2 Borges's interest in and allusion to C. Howard Hinton is most appropriate. Hinton held some strange ideas, among them the notion that the ultimate components of our nervous system are of a higher dimension, thus affording us the capacity to imagine four-dimensional space. The fourth dimension will be discussed in conjunction with relativity theory in the second section of Chapter Four.

3 This quote is relevant to Borges's "New Refutation of Time," which will be the focus of the first section of Chapter Five.

4 Borges was familiar with these arguments from childhood, when his father showed him, "with the aid of a chessboard, the paradoxes of Zeno—Achilles and the tortoise, the unmoving flight of the arrow, the impossibility of motion" (*A,* 207).

5 Borges (*OI,* 110) also describes this problem in referring to the Chinese sophist Hui Tzu, who reasoned that "a stick cut in half each day would be interminable."

6 See Benardete (1964, ch. 6) for numerous variations on this paradox.

7 Borges (*OI,* 114) applies a strange form of this paradox to the problem of knowledge: "To know is to recognize, but it is necessary to have known in order to recognize, but to know is to recognize." See also Borges's reference to Zeno and Kafka's *Castle* (*OI,* 106).

8 Once again, I refer to the "radical meaning variance" theorists, mentioned in note 9 of Chapter One.

9 The problem of this dichotomy between the continuum and the discrete, or the actual and the potentially infinite, is central to quantum

theory. The quantization of energy in quantum theory has led some to speculate upon, and even to the belief of, the microphysical atomiticity of space ("hodons") and time ("chronons"), though few physicists believe in the reality of such today (see Capek 1961 and Whitrow 1969). More commonly one finds the notion of potential infinity beginning with Heisenberg and culminating with the most outspoken advocate of potential infinity, David Bohm (1980). According to Bohm's formulation, space is a continuum, every point of which is potentially the position of, say, an electron, and every instant of continuous time is potentially the time of a physical event. Since Zeno's arrow essentially does not challenge this theory, and if the theory is correct, the idea of granular space and time is no more than wishful thinking (Grünbaum 1967, 111–14).

Notes to Pages: 50–61

10   The notion of tacit and subsidiary or peripheral awareness, in contrast to focal awareness, stems from Polanyi (1958). For treatment of Polanyi's concept from a semiotic perspective commensurate with its use here, see Merrell (1982, 1983, and 1985b).

11   For an excellent, more detailed, study, see Massuh (1980, 138ff).

12   Hayles (1984), whose analysis of Borges will be acknowledged in the following section, has treated this topic in detail.

Chapter Three

1   This reminds us once again of Borges's linear labyrinth at the conclusion of "Death and the Compass." If Zeno's paradox is in force here, and if the line is subdivided infinitely, then the slug from Scharlach's gun will never pierce Lönnrot's chest. But this is finitist thinking. For the finitist's unruly counterpart, in contrast, the shot follows the trajectory we experience for any ordinary projectile—along an infinity of points.

2   Recall Odin's disk, mentioned in the first section of Chapter One, the reverse side of which was invisible.

3   See Rodríguez Monegal (1978, 266–67) on Maurice Blanchot's critique of Borges's inadequate use of the term infinity. However, one must be aware that Borges's infinity is, most properly speaking, an aesthetic model.

4   Cantor inquired into the *how* of infinity, which was considered to be an impossible task during his time. Generally, the accepted opinion had not changed much since Leibniz and Spinoza: the infinite is absolute and therefore incomprehensible. Interestingly, in 1892 Frege predicted that the question of infinity would bring mathematics to the edge of certainty—which actually occurred within a decade (Dauben 1979, 266).

Notes to Pages: 61–67

5 It hardly needs stating that Cantor received mixed reviews. Leopold Kronecker considered him a charlatan and a corrupter of youth; David Hilbert believed he had created a paradise for mathematicians. Russell described him as one of the greatest intellects of the nineteenth century, while Henri Poincaré predicted that later generations would regard set theory as a disease from which mathematics had recovered (Kline 1980, 203). It is worthy of mention also that according to Derrida, a formal structure, motivating Western thought in general, is the "metaphysics of presence," or "logocentrism," which among other things, entails the notion that "[t]he whole is implicated, in the speculative mode of reflection and expression, in each part" (Derrida 1972, xiv). Apparently Derrida would be critical of Cantor, in form if not in content, for his repeating the general equivocations of Western metaphysics. On the other hand, Borges's vindication rests in that he does not exactly adopt Cantorian sets but uses them to show the futility of all absolutist forms of reasoning.

6 Hayles (1984, 158–61) cleverly relates Cantor to some of Borges's works, especially "The Aleph." I have consequently minimized somewhat Cantor's relevance to Borges insofar as it overlaps with her study while acknowledging my debt to her whenever necessary.

7 Hayles (1984, 154) rightly observes that Borges does not simply report Cantor's theorem but subjects it to what Harold Bloom calls a "strong reading," distorting it in such a way that the underlying meaning of his fictions can be revealed through the interplay between the types of infinities.

8 Matte Blanco takes the terms "symmetry" and "asymmetry" directly from mathematics.

9 The "hologram," to be discussed in the fourth section of Chapter Six, is a product of laser photography capable of producing the facsimile of a three-dimensional object on a two-dimensional print (see also Merrell [1991], where I discuss the "hologram" with respect to Peirce's theory of the sign).

10 The "Chinese box" regression will be discussed further in the fifth section of Chapter Six.

11 See, for a layperson's account, Bernstein (1978, 246–65), Bronowski (1966). Also, for a discussion in light of critical and semiotic theory, see Merrell (1985a and 1985b).

12 DeLong (1971, 190) relates this quote to authors of what he calls the "limitative theorems," Gödel, Alonzo Church, T. Skolem, Alfred Tarski, and others.

13 It must be mentioned that I construe Gödel in a metaphorical sense when relating him to Borges, or perhaps it could be said that the opposite is the case.

Notes to Pages: 68–77

14 However, see Sawnor (1972) and Camurati (1987) for notable exceptions.

15 In the preface to *Labyrinths* (*L,* xi), André Maurois refers to Dunne as "the author of such curious books about time." Certainly Borges remained intrigued with Dunne's thesis. What seems to be ignored by most critics is that the thesis, in light of twentieth-century relativity and quantum theory, is not as bizarre as they tend to believe. Dunne, in fact, enjoys a list of rather eminent predecessors. It was mentioned above that at the turn of the century, the notion of a conceivable fourth-dimension and the infinite regress were widely discussed in intellectual circles. To note briefly in passing, Hinton's book, *What Is the Fourth Dimension?* (1887), which influenced Borges according to Barrenechea (1965, 104–6), is referred to repeatedly by Dunne. In *Appearance and Reality* (1897, ch. 3), Bradley demonstrates an infinite regress in our everyday language use. We ordinarily believe the universe to be made up of individuals, a, b, c, ... n that stand in a set of relations to one another. I can say a is to the left of b, and c is to the right of it. These relations "to the left of" and "to the right of" are conceived by Bradley to be higher-order objects that can in turn stand in higher-order relations to other relations and objects. And so on. There is no stopping point. Similarly, Josiah Royce (1901, 504–7), who discusses Bradley—both of whom are cited occasionally by Borges—reasons that one's image of one's own mind is itself an item present to the mind, so the image includes an image that includes an image.

16 Dunne's formulation is not a far cry from the "Minkowski diagram" (to be discussed in the second section of Chapter 4) that models Einstein's space-time continuum, though there is one crucial difference. The "Minkowski diagram" is that of the universe "in bloc," where everything, past, present, and future, exists simultaneously. Dunne's scheme animates this "block," beginning with the "worldline" of an observer in space-time and implying that its consciousness "moves along" this line—an often committed error (Park 1972). Thus the train of thought of Dunne's postulated individual consciousness can be repeated such that it eventually realizes an infinite regress both of time and consciousness.

17 Compare Dunne's formulation to Matte Blanco's "symmetrical being."

Notes to Pages: 77–94

18   Interestingly, with respect to our overcoming death, an argument not unlike Dunne's infinitely regressive explanation is to be found in a novel by Arthur Schnitzler, *Flight into Darkness* (1931, 29–31):

> Leinbach had discovered a proof that there really is no death. It is beyond question, he had declared, that not only the drowning, but all the dying, live over again their whole past lives in the last moment, with a rapidity inconceivable to us others. This remembered life must also have a last moment, and this last moment its own last moment, and so on; hence, dying has itself Eternity: in accordance with the theory of limits one approached death, but never got there.

19   Funes's "perceptual grasp," it bears mentioning, is analogous to the problematics of Laplace's hypothetical Superobserver, who, if cognizant of the state of the entire universe at an instant, could presumably determine, by laws of cause and effect, all past and future states. Relativity physics took a giant step toward abolishing this mechanistic ideal, though the final break awaited quantum theory.

20   Alex Comfort (1984, 61–86) has admirably worked this problem out, following, in addition to physicist David Bohm, G. Spencer-Brown, and in general the "new physics." Comfort's formulation bears a rough similarity to Dunne's scheme, save the temporalizing of the "Minkowski diagram."

21   The demise of totalizing perspectives has been shoved to the fore in recent debates regarding postmodernism (Connor 1989; Harvey 1989; Kaplan 1988; Kellner 1989; Lyotard 1984), and antifoundationalism, postempiricism, and postanalytic philosophy (from various disciplines, see R. Bernstein 1983; Clifford and Marcus 1986; Griffin 1988; Poster 1989; Rajchman and West 1985; Rochberg-Halton 1986; Rorty 1979). See Alazraki (1988) for a view of Borges as submerged in modernism but at the same time projecting postmodernist tendencies.

Interlude

1   See also, in this regard, Bunge (1987).

2   Derrida's (1970) inclination toward a freewheeling Dionysian play that lends itself to frenzy, imaginative vision, and above all, hallucination, leads one to conjecture that his "white mythology" (Derrida, 1974b) might also be roughly applicable to science as, in the beginning, a collectively "controlled hallucination," which, over time has become "embedded" within the somnambulist activities of the general scientific community such that the control has been lost. (For further discussion of such "embedment," see Merrell 1982).

3   See Malcolm (1958), on which the following is based.

4   Though I cannot develop this topic further here, given obvious limitations of time and space, the point is adequately made evident elsewhere, in part inspired by Derrida, on the philosophy of history as narrative (Danto 1985) and philosophy as a "kind of writing" (Rorty 1982).

5   It bears mentioning, though pursuit of the problem lies beyond the scope of this study, that unfortunately, Lyotard and hermeneuticists in general usually direct their attack against an outmoded science as it is conceived in the social and human sciences (see Toulmin 1982; Weinsheimer 1985). In contrast, I speak here of the physical sciences, and most specifically of physics (see also Merrell 1988).

Chapter Four

1   See, in general, the essays on Einstein in Holton (1973, ch. 2).

2   See also Redekop (1980), who relates Borges's finite and unbounded labyrinths to the work of Escher.

3   In the words of James E. Irby in the introduction to *Other Inquisitions* (1964, xiv): "Borges' world-pictures all seem to join in postulating that the world is a supreme mind about to emerge from its symbols and reveal the unity of all things *sub specie aeternitatis*." Later, however, Irby says Borges does not believe this, for chaos always threatens. I would agree in part. Borges believed that such a vantage point is not possible; nevertheless, his game of striving never ceased.

4   Interestingly, Rucker (1977, 130) had an experience with Hinton's mental exercises, but, he remarks, they so frightened him that he gave them up.

5   See, for example, *DT,* 88; *L,* 14 and 90; *BS,* 51.

6   See Dobbs (1972) on four-dimensional perception and Comfort (1984, ch. 2) on the fourth dimension as a mental act.

7   This discussion refers to Abbott's *Flatland* (1952) and Burger's *Sphereland* (1968), which are also the focus of many of Rucker's (1977, 1983, and 1984) descriptions.

8   See Rucker (1984) on the idea of space-time for the layperson and Comfort (1984) for a more abstract presentation.

9   Though many doubters existed when Einstein first announced his theory, it gradually gained acceptance as physicists were attracted to its elegance and convinced by the accumulation of empirical evidence, especially after an experiment by Arthur Eddington demonstrating

Notes to Pages: 131–42

that the sun's gravity actually bends the light from a distant star before it reaches the earth. David Layzer (1990, 5) tells us that today, most physicists would tend to agree with Einstein as well as, surprisingly, with Laplace's Superobserver, that their world is a "block" universe. However, dissenters exist, and their number had been growing of recent. One of the foremost critics of the timeless "block" universe is Ilya Prigogine, who hypothesizes an irreversible, self-organizing universe in constant evolution. Prigogine's "god" is no archivist revealing an infinite sequence he/she designed once and for all time but continues the labor of creation as the universe unfolds (see Prigogine and Stengers 1984).

10 Note that this use of disks implies a binary system such as that of the computer or the *I Ching*.

Chapter Five

1 There are also numerous references to time in this sense in Borges's earlier essays. To cite only two examples, in *Historia de la eternidad* (*HE*, 11) he refers to a passage in Plotinus's *Eneadas* that maintains that to define the nature of time it is imperative that one previously know eternity, and in *Discusión* (*D*, 142), that the chronology of "Bovard et Pecuchet" was examined and proved. The book is replete with circumstances and events, but time, nonetheless, stands still.

2 Borges, strangely enough, often refers to eternity, which seems generally to lie outside the vocabulary of most other twentieth-century philosophical writings about time. If "time" is circular, or simultaneous, what is the nature of eternity? Borges reasons that perhaps it is merely an infinite prolongation of time, but a separate reality—and, surprisingly, this does not appear to worry him unduly. Time, he says, is "a tremulous and exigent problem for us, perhaps the most vital problem of metaphysics; eternity is a game, a worn-out hope" (*HE*, 11). Borges's personal concept of eternity is devoid of God, or of any other owner, for that matter. Nor is it populated with static archetypes. Elsewhere, Borges mentions that there are basically two concepts of eternity: that of the realist, with his universe of archetypes, and that of the nominalist, who desires to coagulate all details of the universe in an instant. Once again, also, Borges does not allow a decided advantage to either of the two concepts; more appropriately, he continues to oscillate between them.

3 See also Rucker (1983, 120–21) who briefly relates Gödel to Borges. Rucker (1984, 218–19) also refers to the problem with Dunne and the "animated block."

4   Here, as elsewhere in this inquiry, it is quite conceivable that the deconstructionist, following Derrida, would charge, albeit rather indirectly, that Einstein's "block" universe suffers from a "logocentric" bias. Be that as it may, full consideration of this intriguing topic would entail a separate, and I would imagine an exceedingly complex, study.

5   Hayles (1984, 158–59) remarks that the world can be divided up and rearranged into a sphere the size of a ball. She mentions that the golf ball would be a good example of the Aleph's size. Actually, it could as easily have been the size of a marble, a buckshot, or a spore.

6   Ken Wilber (1984), in the introduction to *Quantum Questions,* a collection of excerpts from the writings of leading twentieth-century physicists revealing their religious and/or mystical bent, argues convincingly for complementarity rather than compatibility between the scientific view and the mystical view.

7   See Wheelock (1969, 35–36) who, as mentioned, also sets up the complementarity between the Aleph and the Zahir, but for a different purpose: "[T]he obsession with the Zahir is a prototype of the inability to break up and reorganize reality, and . . . Funes had the opposite maladjustment, an incapacity to form abstractions. The Aleph and the Zahir represent the two conceivable forms of understanding: the transcending vision and the meaningful perspective; Funes represents the conceptual duplication of the universe down to the last detail. Borges sees all three of these avenues to reality as illusions."

8   Bell-Villada (1981, 30), in reference to Borges, evokes one of Schopenhauer's favorite parables from the Upanishads, where a student experiences all the beings of the world parading by him in succession, and as each files past the utterance "Tat twam asi" ("This thou art") is heard. A self is simultaneously all selves.

9   Perhaps, as a note, further commentary on "The Other" is in order, where the young Borges who once sat on a bench in Geneva near the Rhone now sits next to the aged Borges in Cambridge facing the Charles River. Regarding this story, Borges reveals in his afterword to the *Book of Sand* (*BS,* 123) that he had to ensure that the two interlocutors were adequately different to be two persons and similar enough to be one. True to form, the younger Borges remarks that "we resemble each other, but you're much older and your hair is gray" (*BS,* 12). And later, the older Borges reflects: "We were too similar and two unalike. We were unable to take each other in, which makes conversation difficult. Each of us was a caricature of the other" (*BS,* 18). They are the same but different; they are different but the same: unity in

Notes to Pages: 156–58

diversity and diversity through unity. This polarity, or complementarity, bears on the identity-diversity of Schrödinger's notion of consciousness, which is, once again, the pattern of struggle enacted between Aleph-Zahir.

Chapter Six

1   See Matson (1964, ch. 4) for a well-documented layperson's account. Arthur Fine (1986, 1–2) observes that in Borges's "The Flowering of an Art" (*BD*, 77–82), H. Bustos Domecq once Quixotesquely marched off in defense of modernism, ultimately recounting the development of Uninhabitables, an architectural movement that began with functionalism and culminated in "Verdussen's" masterly "House of Doors and Windows." The principle underlying this masterpiece of the absurd is the schizoid idea of using all the basic items of habitable dwellings—doors, windows, roofs, walls, etc.—while abandoning the accustomed connections between them. Fine goes on to note that "where we find Don Bustos describing the Uninhabitables as growing out of functionalism, so we see [Niels] Bohr, in parallel, describing the quantum theory as a natural generalization of classical physics. Following Borges's example, we would expect the quantum theory to be a monstrous Uninhabitable, which is exactly what Einstein, perhaps its foremost critic, found it to be."

2   As I have mentioned, Borges claimed to know very little about contemporary physics. In his review of M. Davison's *Free Will Controversy* (*D*, 174–77), he referred to William James, who conjectured that the universe follows a general plan, but the details for executing this plan are left to the actors—us. Then in a footnote he suggested, while admitting to ignorance on the matter, that Heisenberg might not be hostile to James's idea. Borges's intimation, as we shall observe, bears more truth than he most likely realized at that moment.

3   Gribbin (1984) gives an excellent survey. What plagued Einstein regarding quantum theory was basically twofold: (1) he could not go along with the idea that the universe is ultimately based on probability (his well-known "God does not play dice" is a succinct statement of this aversion), and (2) he could not accept quantum theory's irremediable observer participation, which abolished the possibility of an observer-independent reality (see Fine 1986).

4   See, for further discussion, Comfort (1984, 61–86).

5   While the human retina is essentially flat, the image produced by the diagram is as if it were a three-dimensional object. However, a solid object is not capable of producing alternative states as is the Necker cube. When the cube switches from one state to another, it appears that the brain maps the reversal of a three-dimensional object

on the two-dimensional plane as if it were in possession of an additional dimension.

6 In an interesting move, and commensurate with Planck's words, Davies (1983, 110) places this observer within the system of Hofstadter's (1979) "strange loops."

7 For Bohr, the problem lies in seeing the entire experimental arrangement without ambiguity or paradox, but since this cannot be accomplished, therefore the classical view of physics becomes an impossibility (see Hooker 1972).

8 Hayles also compares this sequence to Bach's "Endlessly Rising Canon," employed by Hofstadter in his illustration of recursive series.

9 It must be acknowledged that Wigner's "quantum solipsism" is supported by a minority of the physics community, and has been, perhaps unfortunately, embraced with enthusiasm by a few overanxious students of the paranormal. Yet his thesis is not a simplistic brain-mind duality (i.e., the hardware [brain] categorically separated from the software [mind]). Both brain and mind are caught up in a complex tangle, or "strange loop," as Hofstadter terms it, the macromolecular equivalent of the particle-wave dualism. In general, on the speculative mixture of mind and matter, which provides the means for the universe to come into being, see Gribbin (1984, 208–13), Davies (1983, ch. 3), Wheeler (1980), Wheeler and Patton (1977), Wigner (1970, ch. 13), and Young (1972).

10 Of course, the perennial solution of the great idealistic philosophers has been that mind is logically prior to everything else. Physicists, in contrast, have been loath to allow any consideration of mind into their theories. During the past twenty years, however, "cosmology" has taken on increasing respectability, especially among quantum physicists and astrophysicists. The collection of their ideas, having been labeled the "anthropic cosmological principle," offers a means of relating mind and observership directly to the phenomena traditionally encompassed by classical physics. This principle in its various guises restricts the nature of the universe by asserting that intellectual life, or at least life in some general form, selects the actual universe from among the myriad imaginable universes. Without life, indeed, without mind, there is no universe as we know it. Thus ultimate anthropic constraints are dependent upon living entities, which are part of, yet collaborate with, the universe as a self-organizing whole (see especially Barrow 1990; Barrow and Tipler 1988; Skolimowski 1987). I trust it hardly needs mentioning that in the anthropic cosmological principle we can perceive distinct echoes from Borges's Tlön.

11 The impossibility of this instantaneous travel of information implied by the EPR experiment was one of Einstein's many attempts to refute the Copenhagen interpretation and in so doing demolish quantum uncertainty in favor of his quest for a causal universe.

12 See Merrell (1985a), where I argue that the general view of contemporary physics evinces certain parallels with deconstruction.

13 Jeans (1930, 4), speculating on the origin of the universe, evokes an image related to Borges's Library, which deserves complete quotation:

> [I]f the universe goes on for long enough, every conceivable accident is likely to happen in time. It was, I think, Huxley who said that six monkeys, set to strum unintelligently on typewriters for millions of millions of years, would be bound in time to write all the books in the British Museum. If we examined the last page which a particular monkey had typed, and found that it had chanced, in its blind strumming, to type a Shakespeare sonnet, we would rightly regard the occurrence as a remarkable accident, but if we looked through all the millions of pages the monkeys had turned off in untold millions of years, we might be sure of finding a Shakespeare sonnet somewhere amongst them, the product of the blind play of chance.

Of course, the Library's inhabitants nurture faith in an unknown order resulting from some incomprehensible "play of chance."

14 The quote in the Princeton book is from another translation of Borges, in *Ficciones*, p. 11. Borges's quote is accompanied by a quote from W. James: "Actualities seem to float in a wider sea of possibilities from out of which they were chosen: and *somewhere*, indeterminism says, such possibilities exist, and form part of the truth." See also other comments on Borges's passage in Gribbin (1984, 238) and Talbot (1980, 36–37).

15 Though Everett's cosmology is a radical proposal, it is also metaphysical, and like most philosophical models, it has antecedents. It appears, with variations, in Hindu thought, and in the Kabbalah as the "Kingdom of Edom"—parallel realities with no experiential human time line (Comfort 1984, 100).

16 This is also comparable, the reader will recall, to Dunne's infinitely receding consciousness.

17 For example, if you drop a stone in a still pond, ripples will travel outward in concentric circles from the point the stone entered the water. Now simultaneously drop two stones some feet apart and the

two concentric circles will set up an interference pattern. If the crest of one hits the crest of another, together they will produce a wave of twice the height. If one wave collides with another, they will cancel each other out and produce a patch of calm water. The behavior of light waves is analogous. Laser light consisting of waves of the same frequency are of the purest form, somewhat like a perfectly spherical stone dropped vertically into a completely motionless pond. In laser photography, a laser beam is directed toward a half-silvered mirror such that part of the beam passes through, striking a photographic plate, and part is reflected onto the object to be photographed. The light reflected from the object strikes the photographic plate at an angle with that part of the beam which passed directly through the mirror to produce a complex array of interference patterns. When the plate is developed and illuminated with laser light, the object reproduced on the two-dimensional surface appears to possess the three-dimensional characteristics of the original object. Karl Pribram (1971) maintains that in addition to the mind-brain's thinking and communicating with discrete symbols, it acts also, at a deeper level, like a holograph. This deeper level consists of a nonverbal realm beyond actual symbols, a sort of continuous and indeterminate potentiality that can give rise to a determinate set of symbols.

18   Bohm then relates this metaphor to the tracks of an elementary particle on a photographic emulsion from a bubble chamber. The track is to be regarded as no more than an aspect appearing in immediate perception, as in the case of the drops of dye. Quantum theory demands that movement be described discontinuously, in terms of "quantum jumps." The movement of the particles can be none other than the equivalent of a series of "dots," which our immediate perception is accustomed to grasping in terms of a continuous flow. Bohm (1980, 208) gives the example of the static, discrete frames of a movie film, which, when discontinuously passed through a projector some twenty-four times per second, create the illusion of continuity (in this respect see also the studies of Kolers [1972] and Goodman's [1978] use of them in his *Ways of Worldmaking*). This distinction between continuity and discreteness is also illustrated by Bohm's (1980, 201) analogy to music. A musical score is *there* all at once. To the nonmusician it may be meaningless, but the accomplished musician can rapidly imagine, as her eyes move along the score, a series of sounds of a particular style and quality. If she plays the piece on a piano, there is for the listener a sense of unbroken movement as the notes merge into one another. At a given moment, a certain note is played, but past notes are still reverberating in her consciousness. The simultaneous presence of all these reverberations is necessary for the immediately sensed movement, flow, continuity, for to perceive the notes separated by a greater length of time, the lack of reverberations can destroy the sense of the unbroken whole. This unbroken movement, then, requires a retention of

Notes to Pages: 187–88

past moments, which become more vague and fuzzy as they become further removed from the present, but it also entails projection into the future in terms of expectations. The knowledgeable musician, on perceiving a symphony for the first time, will, after listening to the initial movement, have developed a hypothetical set of expectations regarding the remainder of the piece. These expectations may be verified, and if not, the hypothesis can be altered somewhat and another set of expectations developed, all this transpiring at both conscious and unconscious levels (for the notion of retention, as well as protention, see Husserl [1964]; for expectations in music perception, Meyer [1956]).

19 Bohm (1980, 198–201) suggests that there is a similarity between the order of such immediate experience and the implicate order as it is apprehended by an abstractive act of mind. This reveals the possibility of a coherent mode of understanding the immediate experience of nature in terms of our thought (in effect thus resolving Zeno's arrow paradox). For example, consider how motion is thought of in terms of points along a line. At a certain time, t_1, a particle is at a position x_1, and at a later time t_2, it is at position x_2. The formula for expressing the velocity between the two points on a graph representing it will be static, as if to say "now it is here, now it is there." There is no sense of unbroken wholeness, of the experience of movement. Calculus solves the problem differently. The time interval, t_1-t_2, and the change in position, x_1-x_2, become infinitesimal, and the velocity of the particle is defined as the limit of the ratio of the change in position divided by the change in time ($\Delta x/\Delta t$), as the latter approaches zero. Some reflection reveals that this procedure is as abstract as the previous, for one has no immediate experience of a time interval of zero duration, nor is it possible to see in terms of reflective thought what this could mean. Moreover, calculus entails the notion of continuous movement, but the quantum level movement is discontinuous, so its application is limited to classical concepts (i.e., Bohm's explicate order, such as the movement of billiard balls), which provides an adequate approximation. Movement in the implicate order, on the other hand, does not involve this problem; its movement is "a series of inter-penetrating and intermingling elements in different degrees of unfoldment *all present together*" (Bohm 1980, 203). This activity depends upon the whole enfolded order, which is continuous and determined by the relationships of copresent elements.

20 The reversibility of this branching model also relates to Borges's "New Refutation." Butler (1973, 154) contends that Borges's essay is a "consciously futile attempt" to negate transitoriness in his effort to affirm eternity. Borges's affirmation of eternity stems, of course, from his "feeling in death" experience. Intuiting that this is the focal point of the essay, rather than the prior disquisition on

idealist metaphysics, Butler suggests further that the essay is written backward, like the branching tree diagram in figure 11 "read" from right to left. Be this as it may, it remains that the essay can also be read from left to right, and Borges's conclusion still inheres, because the essay, *in toto,* remains as a "timeless block," while a reading of it in either direction creates the illusion of time. The two halves of "New Refutation of Time" mirror each other, and each is a microcosm mirroring the web of intertextuality-interconnectedness: the universe as text, or conversely, the text as universe.

Notes to Pages: 188–216

21 See, for general comments, Culler (1976), Kristeva (1969), Morgan (1985), and especially, Worton and Still (1990).

22 Compare this to the static, timeless whole, the Minkowski "block."

23 Lugones and Martínez Estrada are Argentine writers and critics, who, like Borges, have written on "Martín Fierro."

24 It bears mentioning that in addition to Maxwell, Bellone studies Galileo, Kelvin, and Newton.

25 For a brief account, see Gardner (1979, 203–13).

26 Roger Penrose (1989), known for his work on black holes and space-time singularities, demonstrates how it is possible, with his "Penrose tiles," to tile a plane with five-fold symmetry using two shapes: a fat rhombus and a thin rhombus. The remarkable pattern produced is orderly as a whole, though there is no periodicity when considering its parts (compare to the Library). Penrose's tiling pattern visually accounts for "quasi-crystals," which are the equivalent of the above-mentioned aperiodic crystals.

1 One might wish to question the wisdom of adopting this "schizophrenic" view. The Copenhagen interpreter's customary response would be that it is the most viable theory because "it works," and he goes on with the game. Though one might then wish to reprimand him, as do most anti-Copenhagenists, for discontinuing his search for a more general and less multifaceted theory, one must admit that he at least honestly faces up to the fact of his epistemological limitations. Be that as it may, he appears to be able to live in his theoretical world comfortably, especially since there is a ready translation between his two incompatible representations.

Chapter Seven

2 For exposition of and the controversy surrounding the RMV thesis, which, due to limitations of time and space, I cannot develop in

detail here, see Gutting (1980), Lakatos and Musgrave (1970), and Suppe (1977).

3 Popper's work originally appeared in German in 1934.

4 Following Feyerabend, Richard L. Gregory, Escher, Hofstadter, Wittgenstein, and others, I have also referred to this and similar "ambiguous" icons within various discursive contexts (Merrell 1982, 1983, 1985a, and 1985b).

5 Significantly, Bohr once declared, with respect to the language problem in describing quantum events, that we are suspended in language and cannot know which way is up and which is down (in Petersen 1968).

6 I cannot do justice here to the long-standing debate between the Kuhnians and the Popperians concerning whether or not "paradigms" can be transcended. Briefly, for Kuhn, as in general for the RMV theorists, we are caught in the framework of our theories and can step outside solely by conversion to a distinct framework within which we are then held captive. For Popper, this "myth of the framework" is exaggerated; we can break out any time we choose. I would suggest that the plausibility of Kuhn's idea has been exaggerated, for "paradigms" are not totally unbreachable corrals fencing one in: they always reveal, at some juncture, a certain degree of openness. And Popper's freedom of choice is perhaps optimistic, for alternatives are difficult to come by, depending on how high one has climbed up on the belief scale.

7 Feyerabend (1975, 273n) briefly discusses Korner's solution.

8 See, for example, Bloor (1983), Taylor (1971), and Winch (1958); also comments in Bernstein (1983, 109–69).

9 Paracelsus, the astrologist-healer, followed the slogan, "The patients are your textbook, the sickbed is your study." He once threw the medieval book of medicine in a fire and replaced it with the "book of nature," the "doctrine of signatures."

10 Galileo's "gaze," admittedly, is the product of Western thought currently under criticism, or "deconstruction," as it were. I have argued elsewhere (Merrell, 1985a), following poststructuralist principles as well as postempiricist philosophy of science, that the subject's presumed neutral observation of the object is a false doctrine, patterning the erroneous notion that mind is the "mirror of nature." "Nature" is not immediately accessible to "mind," but only mediately so, and there

is no "presence," in all its plenitude, nor is the object identical with itself. There can be no more than *differences*, for one cannot read the same book (universe) twice. This, however, should not eliminate the book-world metaphor, which, more appropriately in poststructuralist jargon, would be text-world (Derrida 1978, 294–300).

11 See Massuh (1980, 161–85) for an excellent detailed study.

12 Borges (*D*, 164) mentions Mauthner's *Diccionario de la filosofía* as one of the books he has most often read.

13 The Library image is a coup de grace to the notion that the universe is ultimately intelligible and describable. For a more condensed image, one need only turn to the Library's solitary book with an infinity of pages, or to Layzer's infinite deck of cards, described at the end of Chapter Four, in an attempt to fathom the Cosmic Knot's complexity.

14 The allusion here is to the "literature of exhaustion," from Stark's (1974) book by way of an essay by John Barth (1974).

Works Cited

By Borges: In Chronological Order

Inquisiciones. Buenos Aires: Editorial Proa, 1925.
El idioma de los argentinos. Buenos Aires: Manuel Gleizer, 1928.
Discusión. Buenos Aires: Emecé, 1957. Originally published with Manuel Gleizer, 1932.
Historia de la eternidad. Buenos Aires: Emecé, 1953. Originally published with Editorial Viau y Zona, 1936.
"La biblioteca total." *Sur* 59 (1939):13–16.
Ficciones. Buenos Aires: Emecé, 1956. Originally published with Editorial Sur, 1944.
Ficciones. New York: Grove Press, 1962. Original title from which most, but not all, of the stories were taken, *Ficciones.* Buenos Aires: Editorial Sur, 1944.
Labyrinths, Selected Stories and Other Writings. Edited by D. A. Yates and J. E. Irby. New York: New Directions, 1962. Selections from *Ficciones.* Buenos Aires: Editorial Sur, 1944; *El Aleph.* Buenos Aires: Losada, 1949; *Discusión.* Buenos Aires: Manuel Gleizer, 1932; *Otras inquisiciones, 1937–1952.* Buenos Aires: Editorial Sur, 1952; *El hacedor.* Buenos Aires: Emecé, 1960.
Other Inquisitions, 1937–1952. Translated by R. L. C. Simms. Austin: University of Texas Press, 1964. Original title, *Otras inquisiciones, 1937–1952.* Buenos Aires: Editorial Sur, 1952.
Dreamtigers. Translated by M. Boyer and H. Morland. Austin: University of Texas Press, 1964. Original title, *El hacedor.* Buenos Aires: Emecé, 1960.

Works Cited

Cuentos breves y extraordinarios, in collaboration with Adolfo Bioy Casares. Buenos Aires: Santiago Rueda, 1967.
"Preface." In *The Narrow Act: Borges' Art of Allusion*, by Ronald Christ. New York: New York University Press, 1969.
The Aleph and Other Stories, 1933–1969. Translated by N. T. di Giovanni. New York: E. P. Dutton, 1970. Most, but not all of the stories from *El Aleph*. Buenos Aires: Editorial Losada, 1949.
Obra poética. Madrid: Alianza Editorial, 1972.
Chronicles of Bustos Domecq. Translated by N. T. di Giovanni. New York: E. P. Dutton, 1976. Original title, *Crónicas de Bustos Domecq.* Buenos Aires: Losada, 1967.
The Book of Imaginary Beings. Translated by N. T. di Giovanni. New York: E. P. Dutton, 1978. Original title, *El libro de los seres imaginarios.* Buenos Aires: Editorial Kier, 1967.
The Book of Sand. Translated by N. T. di Giovanni. New York: E. P. Dutton, 1978. Original title, *El libro de arena.* Buenos Aires: Emecé, 1975.
Siete noches. Mexico: Fondo de Cultura Económica, 1980.

Other References

Abbott, Edwin A.
1952. *Flatland.* New York: Dover Publications.

Agassi, Joseph
1975. *Science in Flux.* Dordrecht, Holland: D. Reidel.

Agheana, Ion T.
1984. *The Prose of Jorge Luis Borges: Existentialism and the Dynamics of Surprise.* American University Studies, Series 2, Romance Languages and Literature, v. 13. New York: Peter Lang.

Alazraki, Jaime
1971. "Tlön y Asterion: anverso y reverso de una epistemología." *Nueva Narrativa Hispanoamericana* 1 (no. 2): 21–32.
1988. "Borges: Entre la modernidad y la postmodernidad." *Revista Hispánica Moderna* 14 (no. 2): 175–79.

Alifano, Robert
1984. *Twenty-Four Conversations with Borges.* Housatonic, Mass.: Lascaux Publishers.

Amaral, Pedro V.
1971. "Borges, Babel y las matemáticas." *Revista Iberoamericana* no. 75:421–28.

Works Cited

Anderson, Robert Roland
    1969. "Jorge Luis Borges and the Circle of Time." *Revista de Estudios Hispánicos* 3:313–18.

Arieti, Silvano
    1976. *Creativity: The Magical Synthesis.* New York: Harper & Row.

Augros, Robert M., and George N. Stanciu
    1984. *The New Story of Science.* Lake Bluff, Ill.: Regnery Gateway.

Bagby, Albert I.
    1965. "The Concept of Time of Jorge Luis Borges." *Romance Notes* 6:99–105.

Barrenechea, Ana María
    1965. *Borges: The Labyrinth Maker.* Translated by R. Lima. New York: New York University Press.

Barrow, John D.
    1990. *The World within the World.* New York: Oxford University Press.

Barrow, John D., and Frank J. Tipler
    1988. *The Anthropic Cosmological Principle.* New York: Oxford University Press.

Barth, John
    1974. "Jorge Luis Borges." In *The Literature of Exhaustion,* by J. Stark, 1–61. Durham, N.C.: Duke University Press.

Bateson, Gregory
    1972. *Steps to an Ecology of Mind.* New York: Ballantine Books.

Beauregard, Olivier Costa de
    1981. "Time in Relativity Theory: Arguments for a Philosophy of Being." In *Voices of Time,* edited by J. T. Fraser, 417–33. Amherst: University of Massachusetts Press.

Belitt, Ben
    1972. "The Enigmatic Predicament: Some Parables of Kafka and Borges." *TriQuarterly* 25:268–93.

Works Cited

Bell-Villada, Gene H.
1981. *Borges and His Fiction: A Guide to His Mind and Art.* Chapel Hill: University of North Carolina Press.

Bellone, Enrico
1980. *A World on Paper: Studies on the Second Scientific Revolution.* Translated by M. & R. Giaccone. Cambridge, Mass.: MIT Press.

Benardete, José A.
1964. *Infinity: An Essay in Metaphysics.* Oxford: Clarendon Press.

Bergson, Henri
1964. *Creative Evolution.* London: Macmillan.

Bernstein, Jeremy
1978. *Experiencing Science.* New York: Basic Books.
1982. *Science Observed: Essays Out of My Mind.* New York: Basic Books.

Bernstein, Richard
1983. *Beyond Objectivity and Relativism: Science, Hermeneutics, and Praxis.* Philadelphia: University of Pennsylvania Press.

Birkhoff, George, and David Birkhoff
1932. "A Mathematical Theory of Aesthetics and Its Application to Poetry and Music." *Rice Institute Pamphlet* 19:89–342.

Blanchot, Maurice
1959. "L'infini littéraire: L'Aleph." In *Le livre à venir,* by M. Blanchot, 116–19. Paris: Gallimard.

Bloor, David
1976. *Knowledge and Social Imagery.* London: Routledge & Kegan Paul.
1983. *Wittgenstein: A Social Theory of Knowledge.* New York: Columbia University Press.

Bohm, David
1957. *Causality and Chance in Modern Physics.* Philadelphia: University of Pennsylvania Press.
1965. *The Special Theory of Relativity.* New York: W. A. Benjamin.
1980. *Wholeness and the Implicate Order.* London: Routledge & Kegan Paul.

Works Cited

Bohr, Niels
 1934. *Atomic Theory and the Description of Nature.* Cambridge: Cambridge University Press.
 1958. *Atomic Physics and Human Knowledge.* New York: John Wiley.

Bolter, J. David
 1984. *Turing's Man: Western Culture in the Computer Age.* Chapel Hill: University of North Carolina Press.

Bolzano, Bernard
 1950. *Paradoxes of the Infinite.* Translated by Fr. Prihonsky. New Haven, Conn.: Yale University Press.

Born, Max
 1943. *Experiment and Theory of Physics.* Cambridge: Cambridge University Press.

Bradley, F. H.
 1897. *Appearance and Reality.* New York: Macmillan.

Briggs, John, and F. David Peat
 1989. *Turbulent Mirror.* New York: Harper & Row.

Broglie, Louis de
 1954. *The Revolution in Physics.* New York: Noonday Press.
 1955. *Physics and Microphysics.* London: Hutchinson.
 1959. "A General Survey of the Scientific Work of Albert Einstein." In *Einstein: Philosopher-Scientist,* edited by P. A. Schilpp, 107–27. LaSalle, Ill.: Open Court.

Bronowski, Jacob
 1966. "The Logic of the Mind." *American Scientist* 54 (no. 1): 1–14.

Bruner, Jerome
 1957. "Going Beyond the Information Given." In *Contemporary Approaches to Cognition: A Symposium Held at the University of Colorado,* 41–69. Cambridge, Mass.: Harvard University Press.

Buchanan, Scott Milross
 1962. *Poetry and Mathematics.* New York: John Day.

Buckley, Paul, and F. David Peat, eds.
 1979. *A Question of Physics: Conversations in Physics and Biology.* Toronto: University of Toronto Press.

Works Cited

Bunge, Mario
1963. *The Myth of Simplicity: Problems of Scientific Philosophy.* Englewood Cliffs, N.J.: Prentice-Hall.
1987. "Borges y Einstein, o la fantasía en arte y en ciencia." *Revista de Occidente* 73:45–62.

Burger, Dionys
1968. *Sphereland: A Fantasy about Curved Space and an Expanding Universe.* Translated by C. J. Rheinboldt. New York: Thomas Y. Crowell.

Burgin, Richard
1968. *Conversations with Jorge Luis Borges.* New York: Holt, Rinehart & Winston.

Butler, Colin
1973. "Borges and Time: With Particular Reference to 'New Refutation of Time.' " *Orbis Litterarum* 28:148–61.

Campbell, Jeremy
1982. *Grammatical Man: Information, Entropy, Language, and Life.* New York: Simon & Schuster.

Camurati, Mireya
1987. "Borges, Dunne y la regresión infinita." *Revista Iberoamericana* 53 (no. 141): 925–31.

Capek, Milic
1961. *The Philosophical Impact of Contemporary Physics.* New York: American Book Co.

Capra, Fritjof
1975. *The Tao of Physics.* Berkeley, Calif.: Shambhala Publications.

Carroll, Robert C.
1979. "Borges and Bruno: The Geometry of Infinity in 'La muerte y la brújula.' " *Modern Language Notes* 94 (no. 2): 321–42.

Cassirer, Ernst
1953-57. *The Philosophy of Symbolic Forms.* Translated by R. Manheim. 3 vols. New Haven, Conn.: Yale University Press.

Castañeda, Hector Neri
   1979. "Fiction and Reality: Their Fundamental Consequences." *Poetics* 8 (no. 1/2): 31–62.

Charbonnier, Georges
   1967. *Entretiens avec Jorge Luis Borges.* Paris: Gallimard.

Chisholm, Roderick M.
   1973. "Beyond Being and Nonbeing." *Philosophical Studies* 24:245–57.
   1982. *Brentano and Meinong Studies.* Atlantic Highlands, N.J.: Humanities Press.

Christ, Ronald
   1967. "Jorge Luis Borges, an Interview." *Paris Review,* no. 40: 116–64.
   1969. *The Narrow Act: Borges' Art of Allusion.* New York: New York University Press.
   1972. "Borges at NYU." In *Prose for Borges,* edited by C. Newman and M. Kinzie, 396–411. Evanston, Ill.: Northwestern University Press.

Clarke, C. J. S.
   1977. "The Hinterland Between Large and Small." In *The Encyclopedia of Ignorance: Everything You Wanted to Know about the Unknown,* edited by R. Duncan and M. Weston-Smith. New York: Simon & Schuster.

Clifford, James, and George E. Marcus, eds.
   1986. *Writing Culture: The Poetics and Politics of Ethnography.* Berkeley and Los Angeles: University of California Press.

Colodny, Robert G.
   1972. "Introduction." In *Paradigms and Paradoxes: The Philosophical Challenge of the Quantum Domain,* edited by R. G. Colodny, xi–xix. Pittsburgh, Pa.: University of Pittsburgh Press.

Comfort, Alex
   1984. *Reality and Empathy: Physics, Mind, and Science in the 21st Century.* Albany: State University of New York Press.

Connor, Steven
   1989. *Postmodernist Culture: An Introduction to Theories of the Contemporary.* Oxford: Basil Blackwell.

Works Cited

Culler, Jonathan
 1976. "Presupposition and Intertextuality." *Modern Language Notes* 91 (no. 6): 1380–96.

Danto, Arthur C.
 1985. *Narration and Knowledge*. New York: Columbia University Press.

Dantzig, Tobias
 1930. *Number: The Language of Science*. 4th ed. New York: The Free Press.

Dauben, Joseph Warren
 1979. *Georg Cantor: His Mathematics and Philosophy of the Infinite*. Cambridge, Mass.: Harvard University Press.

Davies, Paul C. W.
 1974. *The Physics of Time Asymmetry*. Berkeley and Los Angeles: University of California Press.
 1981. *The Edge of Infinity: Where the Universe Came From and How It Will End*. New York: Simon & Schuster.
 1983. *God and the New Physics*. New York: Simon & Schuster.
 1984. *Superforce: The Search for a Grand Unified Theory of Nature*. New York: Simon & Schuster.

Davies, Paul C. W., and J. Brown, eds.
 1988. *Superstrings: A Theory of Everything*. Cambridge: Cambridge University Press.

Deleuze, Gilles
 1983. *Nietzsche and Philosophy*. Translated by H. Tomlinson. New York: Columbia University Press.

Deleuze, Gilles, and Félix Guattari
 1983. *Anti-Oedipus: Capitalism and Schizophrenia*. Minneapolis: University of Minnesota Press.

DeLong, Howard
 1971. *A Profile of Mathematical Logic*. New York: Addison-Wesley.

de Man, Paul
 1971. *Blindness and Insight: Essays in the Rhetoric of Contemporary Criticism*. Oxford: Oxford University Press.

Dembo, L. S.
    1970. "An Interview with Jorge Luis Borges." *Contemporary Literature* 11 (no. 3): 315–23.

Derrida, Jacques
    1970. "Structure, Sign, and Play in the Discourse of the Human Sciences." In *The Structuralist Controversy: The Language of Criticism and the Sciences of Man*, edited by R. Macksey and E. Donato, 247–65. Baltimore, Md.: Johns Hopkins University Press.
    1972. *Marges de la philosophie*. Paris: Minuit.
    1974a. *Of Grammatology*. Translated by G. C. Spivak. Baltimore, Md.: Johns Hopkins University Press.
    1974b. "White Mythology: Metaphor in the Text of Philosophy." *New Literary History* 6 (no. 1): 5–74.
    1978. *Writing and Difference*. Translated by A. Bass. Chicago, Ill.: University of Chicago Press.

DeWitt, Bryce S.
    1973. "Quantum Mechanics and Reality." In *The Many-Worlds Interpretation of Quantum Mechanics*, edited by B. S. DeWitt and N. Graham. Princeton, N.J.: Princeton University Press.

DeWitt, Bryce S., and Neill Graham, eds.
    1973. *The Many-Worlds Interpretation of Quantum Mechanics*. Princeton, N.J.: Princeton University Press.

Dirac, Paul A. C.
    1963. "The Physicist's Picture of Nature." *Scientific American* 208 (no. 5): 45–53.

Dobbs, H. A. C.
    1972. "The Dimensions of the Sensible Present." In *The Study of Time*, edited by J. T. Fraser, F. C. Haber, and G. H. Müller, 274–92. New York: Springer-Verlag.

Dunne, J. W.
    1927. *An Experiment with Time*. New York: Macmillan.
    1934. *The Serial Universe*. London: Faber & Faber.

Dyson, Freeman
    1958. "Innovation in Physics." *Scientific American* 199 (no. 3): 74–82.

Works Cited

Eco, Umberto
1984. *Semiotics and the Philosophy of Language.* Advances in Semiotics. Bloomington: Indiana University Press.

Eddington, Arthur
1958a. *The Philosophy of Physical Science.* Ann Arbor: University of Michigan Press.
1958b. *The Nature of the Physical World.* Ann Arbor: University of Michigan Press.

Einstein, Albert
1934. *The World As I See It.* New York: Covici Freide.
1944. "Remarks on Bertrand Russell's Theory of Knowledge." In *The Philosophy of Bertrand Russell,* edited by P. A. Schilpp, 278–91. LaSalle, Ill.: Open Court.
1949a. "Autobiographical Notes." In *Albert Einstein: Philosopher-Scientist,* edited by P. A. Schilpp, 2–94. LaSalle, Ill.: Open Court.
1949b. "Reply to Criticisms." In *Albert Einstein: Philosopher-Scientist.* edited by P. A. Schilpp, 665–88. LaSalle, Ill.: Open Court.

Einstein, Albert, and Leopold Infeld
1938. *The Evolution of Physics: The Growth of Ideas from Early Concepts to Relativity and Quanta.* New York: Simon & Schuster.

Eliot, T. S.
1941. *Selected Essays.* London: Faber & Faber.

Escher, Maurits C.
1971. "Approaches to Infinity." In *The World of M. C. Escher,* 15–16. New York: Harry N. Abrams.

d'Espagnat, Bernard
1983. *In Search of Reality.* New York: Springer-Verlag.

Fadiman, Clifton, ed.
1958. *Fantasia Mathematica.* New York: Simon & Schuster.

Ferrer, Manuel
1971. *Borges y la nada.* London: Tamesis Books.

Feyerabend, Paul
1975. *Against Method.* London: NLB.
1978. *Science in a Free Society.* London: NLB.

Works Cited

Findlay, J. N.
1963. *Meinong's Theory of Objects and Values.* Oxford: Clarendon Press.

Fine, Arthur
1986. *The Shaky Game: Einstein, Realism, and the Quantum Theory.* Chicago, Ill.: University of Chicago Press.

Frank, Rosalyn M., and Nancy Vosburg
1977. "Textos y contratextos en 'El jardín de los senderos que se bifurcan.'" *Revista Iberoamericana*, nos. 100–01:517–34.

Fraenkel, Abraham Adolf
1953. *Abstract Set Theory.* Amsterdam: North-Holland.

Foucault, Michel
1970. *The Order of Things.* New York: Pantheon Books.

Galileo
1957. "The Assayer." In *Discoveries and Opinions of Galileo.* Translated by S. Drake. Garden City, N.J.: Doubleday.

Gardner, Martin
1979. *The Ambidextrous Universe: Mirror Asymmetry and Time-Reversed Worlds.* 2nd ed. New York: Charles Scribner's Sons.

Geertz, Clifford
1983. *Local Knowledge: Further Essays in Interpretive Anthropology.* New York: Basic Books.

Gleick, James
1987. *Chaos: Making a New Science.* New York: Viking.

Gödel, Kurt
1949. "A Remark about the Relationship between Relativity Theory and Idealistic Philosophy." In *Albert Einstein: Philosopher-Scientist,* edited by P. A. Schilpp, 557–62. LaSalle, Ill.: Open Court.

Gombrich, E. H.
1960. *Art and Illusion.* Princeton, N.J.: Princeton University Press.

Works Cited

Goodman, Nelson
1970. "Seven Strictures on Similarity." In *Problems and Projects,* by N. Goodman, 437–46. Indianapolis, Ind.: Bobbs-Merrill.
1978. *Ways of Worldmaking.* Indianapolis, Ind.: Hackett Publishing Co.

Green, Martin
1964. *Science and the Shabby Curate of Poetry: Essays about the Two Cultures.* New York: W. W. Norton.

Gribbin, John
1977. *White Holes.* New York: Dell Publishing Co.
1979. *Time-Warps.* New York: Dell Publishing Co.
1984. *In Search of Schrödinger's Cat: Quantum Physics and Reality.* New York: Bantam Books.

Griffin, David Ray, ed.
1988. *The Reenchantment of Science: Postmodern Proposals.* Albany: State University of New York Press.

Grünbaum, Adolph
1950. "Relativity and the Atomicity of Becoming." *The Review of Metaphysics* 4:144–60.
1967. *Modern Science and Zeno's Paradoxes.* Middletown, Conn.: Wesleyan University Press.

Guillen, Michael
1983. *Bridges to Infinity: The Human Side of Mathematics.* Los Angeles, Calif.: Jeremy F. Tarcher.

Gutierrez Girardot, Raul
1959. *Jorge Luis Borges, un ensayo de interpretación.* Madrid: Insula.

Gutting, Gary, ed.
1980. *Paradigms and Revolutions: Applications and Appraisals of Thomas Kuhn's Philosophy of Science.* Notre Dame, Ind.: University of Notre Dame Press.

Hacking, Ian
1985. "Styles of Scientific Reasoning." In *Post-Analytic Philosophy,* edited by J. Rajchman and C. West, 145–65. New York: Columbia University Press.

Works Cited

Hadamard, Jacques
1945. *The Psychology of Invention in the Mathematical Field.* Princeton, N.J.: Princeton University Press.

Hanson, Norwood R.
1958. *Patterns of Discovery.* Cambridge: Cambridge University Press.

Hardy, G. H.
1967. *Mathematician's Apology.* Cambridge: Cambridge University Press.

Harvey, David
1989. *The Condition of Postmodernity: An Enquiry into the Origins of Cultural Change.* Oxford: Basil Blackwell.

Hayles, N. Katherine
1984. "Subversion: Infinite Series and Transfinite Numbers in Borges' Fictions." In *The Cosmic Web: Scientific Field Models and Literary Strategies in the Twentieth Century,* by N. K. Hayles. Ithaca, N.Y.: Cornell University Press.
1990. *Chaos Bound: Orderly Disorder in Contemporary Literature and Science.* Ithaca, N.Y.: Cornell University Press.

Hawking, Stephen W.
1988. *A Brief History of Time: From the Big Bang to Black Holes.* New York: Bantam Books.

Heidegger, Martin
1971. *On the Way to Language.* Translated by P. D. Hertz. New York: Harper & Row.

Heisenberg, Werner
1958a. *Physics and Philosophy.* New York: Harper & Row.
1958b. *The Physicist's Conception of Nature.* Translated by A. J. Pomerans. New York: Harcourt, Brace.
1966. *Philosophical Problems of Nuclear Science.* Greenwich, Conn.: Fawcett.
1972. *Physics and Beyond: Encounters and Conversations.* Translated by A. J. Pomerans. New York: Harper & Row.

Hesse, Mary
1966. *Models and Analogies in Science.* Notre Dame, Ind.: University of Notre Dame Press.

Works Cited

Hinton, C. Howard
1887. *What Is the Fourth Dimension?* London: Sonnenschein.
1888. *The New Era of Thought.* London: Sonnenschein.

Holton, Gerald
1973. *Thematic Origins of Scientific Thought: Kepler to Einstein.* Cambridge, Mass.: Harvard University Press.

Hofstadter, Douglas R.
1979. *Gödel, Escher, Bach: An Eternal Golden Braid.* New York: Basic Books.

Hooker, Clifford A.
1972. "The Nature of Quantum Mechanical Reality: Einstein versus Bohr." In *Paradigms and Paradoxes: The Philosophical Challenge of the Quantum Domain,* edited by R. G. Colodny, 67–302, Pittsburgh, Pa.: University of Pittsburgh Press.

Husserl, Edmund
1964. *The Phenomenology of Internal Time-Consciousness.* Translated by J. S. Churchill. Bloomington: Indiana University Press.

Huxley, Aldous
1963. *Literature and Science.* New York: Harper & Row.

Irby, James E.
1964. "Introduction." In *Other Inquisitions, 1937–1952,* by Jorge Luis Borges. Translated by R. L. C. Simms, ix–xv. Austin: University of Texas Press.

Jaki, Stanley L.
1978. *The Road of Science and the Ways to God.* Chicago, Ill.: University of Chicago Press.

James, Roger S.
1982. *Physics as Metaphor.* New York: New American Library.

Jammer, Max
1962. *Concepts of Force.* New York: Harper & Row.

Jeans, James
1930. *The Mysterious Universe.* New York: E. P. Dutton.
1943. *Physics and Philosophy.* Cambridge: Cambridge University Press.

Works Cited

Kant, Immanuel
 1950. *Prolegomena to Any Future Metaphysics* (first published in 1783). Indianapolis, Ind.: Bobbs-Merrill.

Kaplan, E. Ann, ed.
 1988. *Postmodernism and Its Discontents: Theories, Practices.* London: Verso.

Kellner, Douglas, ed.
 1989. *Postmodernism, Jameson, Critique.* Washington, D.C.: Maisonneuve.

Kline, Morris
 1980. *Mathematics: The Loss of Certainty.* Oxford: Oxford University Press.

Koestler, Arthur
 1963. *The Sleepwalkers: A History of Man's Changing Vision of the Universe.* New York: Grosset & Dunlap.
 1964. *The Act of Creation: A Study of the Conscious and Unconscious in Science and Art.* New York: Dell Publishing Co.

Kolers, Paul A.
 1972. *Aspects of Motion Perception.* Oxford: Pergamon Press.

Körner, Stephen
 1970. *Categorial Frameworks.* Oxford: Basil Blackwell.

Koyré, Alexandre
 1939. *Etudes galiléenes.* Paris: Hermann.

Kristeva, Julia
 1969. *Semiotiké: Recherches pour une sémanalyse.* Paris: Seuil.

Kuhn, Thomas S.
 1970. *The Structure of Scientific Revolutions.* Chicago, Ill.: University of Chicago Press.
 1977. *The Essential Tension: Selected Studies in Scientific Tradition and Change.* Chicago, Ill.: University of Chicago Press.

Kybury, H. E.
 1961. *Probability and the Logic of Rational Belief.* Middletown, Conn.: Wesleyan University Press.

Works Cited

Lakatos, Imre, and Alan Musgrave, eds.
1970. *Criticism and the Growth of Knowledge.* Cambridge: Cambridge University Press.

Layzer, David
1975. "The Arrow of Time." *Scientific American* 233 (no. 6): 56–69.
1990. *Cosmogenesis: The Growth of Order in the Universe.* New York: Oxford University Press.

Lee, T. D.
1966. "Space Inversion, Time Reversal and Particle Antiparticle Conjugation." *Physics Today* 19:23–31.

Lefebve, Maurice-Jean
1964. "La jolie Tristan ou une estétique de l'infini." *La Nouvelle revue française,* no. 67:102–6.

Lehmann, Hugh
1979. *Introduction to the Philosophy of Mathematics.* Totowa, N.J.: Rowman & Littlefield.

Lorich, Bruce
1973. "Borges's Puzzle of Paradoxes." *Southwest Review* 58:53–65.

Lyotard, Jean-François
1984. *The Postmodern Condition: A Report on Knowledge.* Translated by G. Bennington and B. Massumi. Minneapolis: University of Minnesota Press.

M. del Río, Carmen
1978. "Borges' 'Pierre Menard' or Where Is the Text." *Kentucky Romance Quarterly* 25 (no. 4): 459–69.

McCulloch, Warren Sturgis
1965. *Embodiments of Mind.* Cambridge, Mass.: MIT Press.

Macintyre, Alasdair
1980. "Epistemological Crisis, Dramatic Narrative, and the Philosophy of Science." In *Paradigms and Revolutions,* edited by G. Gutting, 54–74. Notre Dame, Ind.: University of Notre Dame Press.

McTaggert, J. M. E.
1927. *The Nature of Existence.* Vol. 2. Cambridge: Cambridge University Press.

Malcolm, Norman
    1959. *Dreaming*. London: Routledge & Kegan Paul.

Malraux, André
    1951. *The Voices of Silence*. New York: Doubleday & Co.

Mandelbrot, Benoit
    1982. *The Fractal Geometry of Nature*. San Francisco, Calif.: Freeman.

Margenau, Henry
    1949. "Einstein's Concept of Reality." In *Albert Einstein: Philosopher-Scientist*, edited by P. A. Schilpp, 245–68. LaSalle Ill.: Open Court.

Massuh, Gabriela
    1980. *Borges: una estética del silencio*. Buenos Aires: Editorial de Belgrano.

Matson, Floyd
    1964. *The Broken Image: Man, Science and Society*. New York: Doubleday & Co.

Matte Blanco, Ignacio
    1975. *The Unconscious as Infinite Sets: An Essay in Bi-Logic*. London: Duckworth.

Maurois, André
    1962. "Preface." In *Labyrinths, Selected Stories and Other Writings*, by Jorge Luis Borges, edited by D. A. Yates and J. E. Irby, ix–xiv. New York: New Directions.

Medewar, Peter
    1982. *Pluto's Republic*. Oxford: Oxford University Press.

Merrell, Floyd
    1980. "Understanding Fictions." *Kodikas/Code: An International Journal of Semiotics* 2 (no. 3): 235–48.
    1982. *Semiotic Foundations: Steps to an Epistemology of Written Texts*. Bloomington: Indiana University Press.
    1983. *Pararealities: The Nature of Our Fictions and How We Know Them*. Purdue University Monographs in Romance Languages. Amsterdam: John Benjamins.
    1985a. *Deconstruction Reframed*. West Lafayette, Ind.: Purdue University Press.
    1985b. *A Semiotic Theory of Texts*. Berlin: Mouton de Gruyter.

Works Cited

1988. "An Uncertain Semiotic." In *The Current in Criticism*, edited by C. Koelb and V. Lokke, 243–64. West Lafayette, Ind.: Purdue University Press.
1991. *Signs Becoming Signs: Our Perfusive, Pervasive Universe.* Bloomington: Indiana University Press (in press).

Meyer, Leonard B.
1956. *Emotion and Meaning in Music.* Chicago, Ill.: University of Chicago Press.
1967. *Music, the Arts, and Ideas: Patterns and Predictions in Twentieth-Century Culture.* Chicago, Ill.: University of Chicago Press.

Mignolo, Walter
1977. "Emergencia, espacio, 'mundos posibles': las propuestas epistemológicas de Jorge L. Borges." *Revista Iberoamericana*, nos. 100–101:357–79.

Milleret, Jean de
1967. *Entretiens avec Jorge Luis Borges.* Paris: Pierre Belfond.

Miller, Arthur I.
1978. "Visualization Lost and Regained: The Genesis of the Quantum Theory in the Period 1913-27." In *On Aesthetics in Science*, edited by J. Wechsler, 73–102. Cambridge, Mass.: MIT Press.
1986. *Imagery in Scientific Thought.* Cambridge, Mass.: MIT Press.

Morgan, Thais E.
1985. "Is There an Intertext in the Text? Literary and Interdisciplinary Approaches to Intertextuality." *American Journal of Semiotics* 3 (no. 4): 1–40.

Morris, Richard
1983. *Dismantling the Universe: The Nature of Scientific Discovery.* New York: Simon & Schuster.

Nietzsche, Friedrich
1913. *The Will to Power.* In *The Complete Works of Friedrich Nietzsche*, vol. 9. Edinburgh: Foulis.

Northrop, Fillmer S. C.
1949. "Einstein's Conception of Science." In *Albert Einstein: Philosopher-Scientist*, edited by P. A. Schilpp, 387–408. LaSalle, Ill.: Open Court.

Works Cited

Oppenheimer, Robert
  1954. *Science and Common Understanding*. New York: Simon & Schuster.

Otto, Rudolf
  1932. *Mysticism East and West*. Translated by B. L. Bracey and R. C. Payne. New York: Collier.

Ouspensky, P. D.
  1922. *Tertium Organum*. New York: Alfred A. Knopf.

Pagels, Heinz R.
  1983. *The Cosmic Code: Quantum Physics as the Language of Nature*. New York: Bantam Books.

Park, David
  1972. "The Myth of the Passage of Time." In *The Study of Time*, edited by J. T. Fraser, F. C. Haber, and G. H. Muller, 110–21. New York: Springer-Verlag.
  1980. *The Image of Eternity: Roots of Time in the Physical World*. New York: New American Library.

Parsons, Terence
  1974. "A Prolegomenon to Meinongian Semantics." *Journal of Philosophy* 61 (no. 16): 561–80.
  1980. *Nonexistent Objects*. New Haven, Conn.: Yale University Press.

Peirce, Charles Sanders
  1960. *Collected Papers*, edited by C. Hartshorne and P. Weiss. Cambridge, Mass.: Harvard University Press, Belknap Press.

Peicovich, Esteban
  1980. *Borges, el palabrista*. Madrid: Letra Viva.

Pérez Gallego, Cándido
  1966. "Borges, o la erudición como fantasía." *Revista de las Indias*, nos. 103–4:107–19.

Petersen, A.
  1968. *Quantum Physics and the Philosophical Tradition*. Cambridge, Mass.: MIT Press.

Penrose, Roger
  1989. *The Emperor's New Mind: Concerning Computers, Minds, and the Laws of Physics*. New York: Oxford University Press.

Works Cited

Phillips, Derek L.
1977. *Wittgenstein and Scientific Knowledge: A Sociological Perspective.* Totowa, N.J.: Rowman & Littlefield.

Piaget, Jean
1971. *Psychology and Epistemology: Towards a Theory of Knowledge.* New York: Viking Press.

Planck, Max
1932. *Where Is Science Going?* London: George Allen & Unwin.
1936. *The Philosophy of Physics.* New York: W. W. Norton.

Poincaré, Henri
1958. *The Value of Science.* New York: Dover.

Polanyi, Michael
1958. *Personal Knowledge.* Chicago, Ill.: University of Chicago Press.

Popper, Karl
1959. *The Logic of Scientific Discovery.* New York: Harper & Row.
1972. *Objective Knowledge.* Oxford: Oxford University Press.

Poster, Mark
1989. *Critical Theory and Poststructuralism: In Search of Context.* Ithaca, N.Y.: Cornell University Press.

Pribram, Karl
1971. *Languages of the Brain: Experimental Paradoxes and Principles in Neuropsychology.* Englewood Cliffs, N.J.: Prentice-Hall.

Prigogine, Ilya
1980. *From Being to Becoming: Time and Complexity in the Physical Sciences.* San Francisco, Calif.: W. H. Freeman.

Prigogine, Ilya, and Isabelle Stengers
1984. *Order Out of Chaos: Man's New Dialogue with Nature.* New York: Bantam Books.

Redekop, Ernest H.
1980. "Labyrinths in Time and Space." *Mosaic* 13 (nos. 3–4): 95–113.

Rescher, Nicholas
1975. *A Theory of Possibility.* Oxford: Basil Blackwell.

Rescher, Nicholas, and Robert Brandom
1979. *The Logic of Inconsistency: A Study of Non-Standard Possible-World Semantics and Ontology.* Totowa, N.J.: Rowman & Littlefield.

Richards, I. A.
1926. *Science and Poetry.* New York: W. W. Norton.

Rajchman, John, and Cornel West, eds.
1985. *Post-Analytic Philosophy.* New York: Columbia University Press.

Rochberg-Halton, Eugene
1986. *Meaning and Modality: Social Theory in the Pragmatic Attitude.* Chicago, Ill.: University of Chicago Press.

Rodríguez Monegal, Emir
1978. *Jorge Luis Borges: A Literary Biography.* New York: E. P. Dutton.

Rorty, Richard
1979. *Philosophy and the Mirror of Nature.* Princeton, N.J.: Princeton University Press.
1982. *Consequences of Pragmatism.* Minneapolis: University of Minnesota Press.

Ross, Waldo
1975. "Borges y el problema de las series infinitas." *Anales de Literatura Hispanoamericana,* no. 4:279–84.

Routley, Richard
1979. "The Semantical Structure of Fictional Discourse." *Poetics* 8 (no. 1/2): 3–30

Royce, Josiah
1901. *The World and the Individual.* New York: Macmillan.

Rucker, Rudolf v. B.
1977. *Geometry, Relativity and the Fourth Dimension.* New York: Dover.

Rucker, Rudy [Rudolf v. B.]
1983. *Infinity and the Mind: The Science and Philosophy of the Infinite.* New York: Bantam Books.
1984. *The Fourth Dimension: Toward a Geometry of Higher Reality.* Boston, Mass.: Houghton Mifflin.

Works Cited

Russell, Bertrand
   1926. *Our Knowledge of the External World*. London: George Allen & Unwin.
   1940. *An Inquiry into Meaning and Truth*. London: George Allen & Unwin.
   1946. *The History of Western Philosophy*. London: George Allen & Unwin.
   1973. *Essays in Analysis*. New York: George Braziller.

Sábato, Ernesto
   1945. "Borges: geometrización de la novela." In *Uno y el universo*, by E. Sábato, 104–10. Buenos Aires: Editorial Sudamericana.

Sachs, Mendel
   1988. *Einstein versus Bohr: The Continuing Controversies in Physics*. LaSalle, Ill.: Open Court.

Sawnor, Edna A.
   1972. "Borges and Bergson." *Cuadernos Americanos* 185 (no. 6): 247–54.

Schnitzler, Arthur
   1931. *Flight into Darkness*. Translated by W. A. Drake. New York: Simon & Schuster.

Schrödinger, Erwin
   1967. *What Is Life?* and *Mind and Matter*. Cambridge: Cambridge University Press.

Shultz, Robert A.
   1979. "Analogues of Argument in Fictional Narrative." *Poetics* 8 (no. 1/2): 231–44.

Sklar, Lawrence
   1985. *Philosophy and Spacetime Physics*. Berkeley and Los Angeles: University of California Press.

Skolimowski, Henryk
   1987. "The Interactive Mind in the Participatory Universe," In *The Real and the Imaginary: A New Approach to Physics*, edited by J. E. Charon, 69–94. New York: Paragon.

Spencer-Brown, G.
   1957. *Probability and Scientific Inference*. London: Longmans, Green and Co.
   1979. *Laws of Form*. New York: E. P. Dutton.

Works Cited

Spengler, Oswald
    1932. *The Decline of the West.* Translated by C. F. Atkinson. New York: Alfred A. Knopf.

Stark, John O.
    1974. *The Literature of Exhaustion: Borges, Nabakov, and Barth.* Durham, N.C.: Duke University Press.

Stebbing, L. Susan
    1958. *Philosophy and the Physicists.* New York: Dover.

Sturrock, John
    1977. *Paper Tigers: The Ideal Fictions of Jorge Luis Borges.* Oxford: Clarendon Press.

Sucre, Guillermo
    1970. "La biografía del infinito." *Eco,* No. 125:466–502.

Suppe, Frederick, ed.
    1977. *The Structure of Scientific Theories.* Urbana: University of Illinois Press.

Sypher, Wylie
    1962. *Loss of the Self in Modern Literature and Art.* New York: Random House.

Talbot, Michael
    1980. *Mysticism and the New Physics.* New York: Bantam Books.

Taylor, Charles
    1971. "Interpretation and the Sciences of Man." *Man* 25:3–51.

Todorov, Tzvetan
    1973. *The Fantastic: A Structural Approach to a Literary Genre.* Translated by R. Howard. Cleveland, Ohio: Case Western Reserve University Press.

Toulmin, Stephen
    1972. *Human Understanding.* Vol. 1. Oxford: Oxford University Press.
    1982. "The Construal of Reality: Criticism in Modern and Postmodern Science." In *The Politics of Interpretation,* edited by W. J. T. Mitchell, 99–117. Chicago, Ill.: University of Chicago Press.

Unamuno, Miguel de
    1954. *The Tragic Sense of Life.* New York: Dover.

Works Cited

Updike, John
- 1965. "The Author as Librarian." *New Yorker* 41 (no. 37): 223–46.

Vaihinger, Hans
- 1924. *The Philosophy of "As If": A System of the Theoretical, Practical and Religious Fictions of Mankind.* Translated by C. K. Ogden. 2nd ed. London: Kegan Paul, Trench, Truber.

Wahl, Jean
- 1964. "Les personnes et l'impersonnel." In *L'Herme,* edited by D. de Boux and J. de Milleret, 257–64. Paris: Lettres Modernes.

Waismann, Friedrich
- 1959. "How I See Philosophy." In *Logical Positivism,* edited by A. J. Ayer, 345–80. Glencoe, Ill.: The Free Press.

Walker, Evan Harris
- 1970. "The Nature of Consciousness." *Mathematical Biosciences* 7:138–78.

Watt, Alan
- 1954. *Myth and Ritual in Christianity.* New York: Thames & Hudson.

Wechsler, Judith, ed.
- 1978. *On Aesthetics in Science.* Cambridge, Mass.: MIT Press.

Weiler, Gershan
- 1970. *Mauthner's Critique of Language.* Cambridge: Cambridge University Press.

Weinsheimer, Joel C.
- 1985. *Gadamer's Hermeneutics: A Reading of* Truth and Method. New Haven, Conn.: Yale University Press.

Weyl, Hermann
- 1949. *Philosophy of Mathematics and Natural Science.* Rev. ed., translated by O. Helmer. New York: Atheneum.
- 1952. *Symmetry.* Princeton, N.J.: Princeton University Press.

Wheeler, John A.
- 1957. "Assessment of Everett's 'Relative State' Formulation of Quantum Theory." *Review of Modern Physics* 29 (no. 3): 463–65.

Works Cited

1980. "Beyond the Black Hole." In *Some Strangeness in the Proportion,* edited by H. Woolf, 341–75. New York: Addison-Wesley.

1982. "Bohr, Einstein, and the Strange Lesson of the Quantum." In *Mind in Nature,* edited by R. Q. Elvee, 1–30. New York: Harper & Row.

Wheeler, John, and C. M. Patton.
1977. "Is Physics Legislated by Cosmogony?" In *The Encyclopedia of Ignorance,* edited by R. Duncan and M. Weston-Smith, 19–35. New York: Simon & Schuster.

Wheelock, Carter
1969. *The Mythmaker: A Study of Motif and Symbol in the Short Stories of Jorge Luis Borge.* Austin: University of Texas Press.

Whittaker, Edmund
1954. "W. R. Hamilton." *Scientific American* 190 (no. 5): 82–87.

Whitehead, Alfred North
1925. *Science and the Modern World.* New York: Macmillan.
1929. *Process and Reality.* New York: Macmillan.

Whitrow, G. J.
1969. *The Natural Philosophy of Time.* New York: Harper & Row.

Wigner, Eugene P.
1965. "Violations of Symmetry in Physics." *Scientific American* 213 (no. 66): 28–42.
1969. "The Unreasonable Effectiveness of Mathematics in the Natural Sciences." In *The Spirit and the Uses of the Mathematical Sciences,* edited by T. L. Saaty and F. J. Weyl, 123–40. New York: McGraw-Hill.
1970. *Symmetries and Reflections: Scientific Essays.* Bloomington: Indiana University Press.

Wilber, Ken
1984. "Introduction: Of Shadows and Symbols." In *Quantum Questions: Mystical Writings of the World's Great Physicists,* edited by K. Wilber, 3–29. Boulder, Colo.: Shambhala.

Wilber, Ken, ed.
1982. *The Holographic Paradigm and Other Paradoxes: Exploring the Leading Edge of Science.* Boulder, Colo.: Shambhala.

Works Cited

Winch, Peter
1958. *The Idea of a Social Science and Its Relation to Philosophy.* London: Routledge & Kegan Paul.

Wittgenstein, Ludwig
1953. *Philosophical Investigations.* Translated by G. E. M. Anscombe. New York: Macmillan.
1956. *Remarks on the Foundations of Mathematics.* Translated by G. E. M. Anscombe. New York: Macmillan.
1970. *Zettel.* Translated by G. E. M. Anscombe. Berkeley and Los Angeles: University of California Press.
1972. *On Certainty.* Translated by D. Paul and G. E. M. Anscombe. New York: Harper & Row.
1974. *Philosophical Grammar,* edited by R. Rhees, translated by A. Kenny. Berkeley and Los Angeles: University of California Press.

Woods, John
1974. *The Logic of Fiction.* The Hague: Mouton.

Worton, Michael, and Judith Still, eds.
1990. *Intertextuality: Theories and Practice.* Manchester: Manchester University Press.

Young, Arthur A.
1972. "Consciousness and Cosmology." In *Consciousness and Reality,* edited by C. Musés and A. M. Young, 151–64. New York: Outerbridge & Lazard.

Zalazar, Daniel E.
1976. *Ensayos de interpretación literaria.* Buenos Aires: Crisol.

Zukav, Gary
1979. *The Dancing Wu Li Masters: An Overview of the New Physics.* New York: William Morrow.

Index

(References to Borges's short stories, poems, and essays are under the "Borges" entry.)

Abstraction: in contemporary physics, 89–90
Agassi, Joseph, 29
Agheana, Ion, 117, 199
Alazraki, Jaime, 13
Aleph-null, 60–62. See also Cantor sets
Aleph-Zahir complementarity, 151–52, 226–27, 255n. See also Bohr
Amaral, Pedro, 117
Anthropic cosmological principle, 257n
Aperiodic crystals, 206
Aristotle, 43, 101, 109, 151, 193, 210; logic, 49; opposed to Platonism, 2, 246n; tragedy and comedy in "Averroes' Search," 223
Aspect, Alain, 168–69. See also Einstein-Podolsky-Rosen (EPR) thought experiment, Paradox
Averroes, 74–76
Axiom of choice, 97–98

Bacon, Francis, 86, 235
Bagby, Albert, 139
Barrenechea, Ana María, 55, 239
Barth, John, 246n, 263n
Barthes, Roland, xiii
Bateson, Gregory, 247n
Baudelaire, Charles Pierre, 107
Beauregard, Olivier Costa de, 141
Beauty: in mathematical equations and scientific theories, 88–90. See also Dirac, Wigner
Beckett, Samuel, xiii, 31, 68, 163, 215, 246n
Belitt, Ben, 16
Bellone, Enrico, 196–97
Bell-Villada, Gene, 84, 99–100, 255n
Benardete, José, 58; "sludge" and the paradox of infinity, 59–60. See also Infinity, Paradox
Bergson, Henri, 48, 105
Berkeley, George, xiii, 4, 13, 15, 26, 29, 94–95, 108, 136, 138, 158, 164–65, 170, 242
Bernstein, Jeremy, xiv
Big bang, 112. See also Cone model of universe, Space-time manifold
Black holes, 134, 147–48. See also Space-time manifold, Space-time singularity
Blake, William, 85
Blanchot, Maurice, 249n
"Block" universe, 122, 123–24, 141, 143, 171, 187, 255n, 261n. See also Minkowski "block," Space-time manifold, Space-time singularity
Bloor, David, 19, 229, 262n
Bohm, David, 189, 192–94, 249n, 252n, 259–60n; enfolded-unfolded and implicate-explicate, 183–88; mind mirrors matter, 185–86;

291

Index

time and Zeno, 260n. *See also* Hologram, Quantum theory, Unbroken wholeness

Bohr, Niels, 86, 105, 157, 161, 256n, 257n; complementarity, 85, 212–13. *See also* Copenhagen interpretation, Quantum theory

Bolter, David: on "computer thought," 10–11

Bolzano, Bernard, 55

Book: metaphor of the universe, 234–35, 241

Boole, George, 207

Borges, Jorge Luis: anxiety over infinity, 55; biography writing, 89; Chinese taxonomy, 22; "feeling in death," his experience of, 136, 141, 181; skepticism, xii; "Aleph, The," 7–9, 38, 49, 62, 114, 119, 181, 186, 226–27, 225n—and the "block" universe, 125; and the fourth dimension, 120–21; as a naked singularity, 145–54; "Approach to al-Mu'tasim, The," 51; "Avatars of the Tortoise," 42, 43, 45, 49, 106; "Averroes' Search," 74–76, 221–28; "Book of Sand, The," 56–57; "Borges and I," 92–93; "Chess," 50; "Circular Ruins, The," 33–38, 50, 63; "Congress, The," 247n; "Cyclical Time," 116; "Death and the Compass," 46–48, 106, 203–5; "Disk, The," 7; "Doctrine of Cycles, The," 113–15; "Dream of Coleridge, The," 51, 191–92; "Emma Zunz," 135, 203, 236; "End, The," 195; "Everything and Nothing," 77–78; "Examination of the Work of Herbert Quain," 182, 190; "Funes the Memorious," 78–80; "Garden of Forking Paths, The," 135, 177–82; "God's Script, The," 65–66, 119, 235; *Historia de la eternidad*, 8, 9, 39, 61, 254n; "House of Asterion, The," 62–63; "Immortal, The," 15, 135, 229–33; "Iron Coin, The," 246n; "Library of Babel, The," 49, 58, 63, 68, 69–71, 76, 106, 108, 109–10, 113, 117, 164–65, 172, 182, 234–35—and the "block" universe, 124; and fractal geometry, 200–201; hexagonal structure, 199–201; three modes of conceiving, 124–32; "Lottery of Babylon, The," 118–19, 172, 173–77; "Mirror and the Mask, The," 235; "New Refutation of Time, A," 77, 81, 134–41, 169, 260–61n; "Nothingness of Personality, The," 162; "Other, The," 93–94, 255–56n; "Pascal's Sphere," 108, 130; "Pierre Menard, Author of the *Quixote*," 172, 194–95, 207–8, 237; "Secret Miracle, The," 41–42, 139; "Sect of the Phoenix, The," 50–51; "Shape of the Sword, The," 202; "South, The," 135; "Theme of the Traitor and the Hero," 202; "Theologians, The," 135; "Three Versions of Judas," 201–2; "Time and J. W. Dunne," 76–77; "Tlön, Uqbar, Orbis Tertius," 25–27, 158–61, 164–66, 240–41; "Undr," 235; "Unicorns, The," 21; "Utopia of a Tired Man," 243–44; "Waiting, The," 95–96; "Warrior and the Captive, The," 202–3; "Zahir, The," 5–8, 41, 49, 81, 119, 226–27—and the "block" universe, 124–25; and the fourth dimension, 120–21

Index

Born, Max, 99, 105
Boyle's law of gases, 170
Bradley, Francis Herbert, 2, 81, 137, 143, 144, 190, 251n
Brandom, Robert, 18, 21, 25, 31
Brentano, Franz C., 23
Broglie, Louis de, 122
Bruner, Jerome, 4
Buddha, 52, 144, 186, 192; and the "block" universe, 124
Bunge, Mario, 248n
Burgin, Richard, 33, 87
Burroughs, William, 215
Butler, Colin, 137, 138, 143

Caillois, Roger, xiii
Cantor, Georg, 30, 53–54, 60, 63, 81, 140, 200; Alephs, 64; Cantor-Dedekind axiom, 141; Cantor sets, 66, 78, 106, 114, 186, 250n. *See also* Aleph-null, "The Aleph," "The House of Asterion," Paradox
Capra, Fritjof, 183
Carroll, Robert, 204
Casares, Adolfo Bioy, 26
Cassirer, Ernst, 4
Castañeda, Carlos, 230
Cervantes de Saavedra, Miguel, 168, 172, 189, 195–96, 237, 238
Chance-necessity, 171. *See also* "The Lottery," Necessity-contingency
Chaos: "The Library," 74; "The Lottery," 173–77; and randomness, 74
Chesterton, Gilbert, xiii, 206
Chinese box, 51, 66, 250n; and serial time, 77
Chomsky: generative-transformational grammar, 164
Christ, Ronald, 5, 196
Chuang Tzu, 137, 143
Church, Alonzo, 31
Coleridge, Samuel, 2, 85, 105

Comfort, Alex, 17, 252n, 253n, 256n
Commonsense: notions of reality, 91–92
Computer thought, 10–11, 14
Conceptual framework, 219–20. *See also* Paradigm
Cone model of universe, 115. *See also* Space-time manifold
Consciousness: infinite series, 80–81; quantum physicists unable to ignore, 166–68; symmetry and asymmetry, 64–65. *See also* Dunne, Matte Blanco, Schrödinger, Wigner
Continuum: dense and nondense, 55
Contrary-to-fact conditional: in scientific discourse and storytelling, 100–101
Controlled: dream, 94–96; hallucination, 95, 104, 106, 181, 252n; related to scientist as "sleepwalker," 92–93
Copenhagen interpretation: of quantum mechanics, 28, 105–6, 157, 183, 241, 258n, 261n. *See also* Bohm, Bohr, Quantum theory
Copernicus, 162
Cortázar, Julio, 248n
Cusa, Nicholas de, 45, 109, 128

Dalí, Salvador, 31, 163
Dantzig, Tobias, 25, 45, 140
Davies, Paul, 111, 169, 257n
Deconstruction, 262n
Dedekind, Richard, 140–41; Dedekind cut, 140. *See also* Cantor, Continuum
Deleuze, Gilles, xiii, 180
DeLong, Howard, 250n
de Man, Paul, 236
Democritus, 53, 183
Derrida, Jacques, xiii, 31, 105, 163, 209, 237, 250n, 252n, 255n

293

Index

Descartes, René, xiii, 8, 88, 207, 242
DeWitt, Bryce, 179, 180
Diogenes, 158
Dirac, P. A. C., 88, 90, 212
Dissipative structures, 246n. *See also* Prigogine
Dobbs, H. A. C., 127, 217, 253n
Domecq, H. Bustos, 256n
Duchamp, Marcel, 31, 217–18
Dunne, J. W., 13, 82, 139, 251n, 258n; serial consciousness, 81; serial observer, 76–77, 79
Dyson, Freeman, 161

Eco, Umberto, 18
Eddington, Arthur, 5, 11–12, 14, 15, 44, 85, 150, 162, 163, 231, 236, 253–54n
Eight-fold way, 205. *See also* Quantum theory
Einstein, Albert, 11–12, 27, 28, 29–30, 45, 85, 86, 87, 88, 95, 100, 102, 105, 107, 112, 117, 119, 134, 142, 144, 146–54, 218–19, 241, 253n, 256n. *See also* Relativity theory
Einstein-Podolsky-Rosen (EPR) thought experiment, 168–69
Eliade, Mircea, 198
Eliot, T. S., 189
Enantiomorphs, 125, 198. *See also* Symmetry
Enfolded-unfolded, 183–88. *See also* Bohm, Implicate-explicate
Ensemble-history, 171. *See also* Chance-necessity, "The Lottery," Necessity-contingency
Entropy, 115, 130. *See also* Second law of thermodynamics
Escher, Mavrits C., 31, 33, 71–72, 109, 112, 163, 176–77, 187, 262n
Eternal return, 113, 114, 115–17. See also *Historia de la eternidad*

"Ether," 85
Event horizon, 147. *See also* Space-time singularity
Everett, Hugh, III, 177, 179, 180, 182, 187, 192–93, 258n. *See also* "The Garden," Many-worlds interpretation of quantum mechanics, Quantum theory
Everything-nothing dialectic, 193–94. *See also* "Everything and Nothing"
Experience near-experience distant, 226–28. *See also* Geertz

Faraday, Michael, 84, 196–97
Faustian man, 117, 244
Fernández, Macedonio, xiii
Ferrer, Manuel, 16
Feyerabend, Paul, 45, 86, 101, 214, 216, 217, 218, 220, 224, 227, 229, 238, 239, 247n, 262n. *See also* Incommensurability thesis, Radical meaning variance theory
Fiction, 9; relation to "reality," 1, 15, 16–19, 21, 22–23, 24, 26–27; and dream in "The Circular Ruins," 35–37. *See also* Controlled dream, Controlled hallucination, Hyper-fictionalization
Field theory, 183. *See also* Relativity theory
Fierro, Martín, 195
Flatlander: in three-dimensional universe, 187–88, 232–33. *See also* Möbius strip, Necker cube
Formenter Prize: won by Borges and Beckett, xiii
Foucault, Michel, 22
Fractal geometry, 200–201. *See also* Mandelbrot
Frank, Rosalyn, 188
Frege, Gottlob, 14, 19
Freud, Sigmund, xiii

Gadamer, Hans-Georg, 105
Galilei, Galileo, 86, 101, 131, 196, 234, 235, 262–63n; computer thought, 10–11
Geertz, Clifford: experience near-experience distant, 226–28
Genette, Gérard, 189
Girardot, Raul Gutiérrez, 12
Gödel, Kurt, 31, 54, 59, 67, 68, 71, 81, 142, 143, 144, 189, 206; incompleteness theorems, 151–52; strategy for his proof, 68–69. See also Paradox: liar
Gombrich, E. H., 4, 215
Gómez de Quevedo, Francisco, xiii
Góngora, Luis de, xiii
Goodman, Nelson, 3–5, 9, 13, 22, 23, 26, 30, 45, 195
Grand unified theory, 70, 246n. See also Relativity theory
Greeks: regarding computer thought, 10; Eleatics' mind-dependent formulations, 44–45; fear of infinity, 54
Gregory, Richard L., 262n
Gribbin, John, 107, 256n
Guattari, Félix, 180

Hamilton, William Rowe, 14, 98–99, 197. See also Quaternions
Hamlet, 70, 118, 166, 192
Hanson, Norwood, 45, 216, 247n
Hardy, G. H., 241
Hawking, Stephen, 246n
Hayles, N. Katherine, 70–71, 81, 97–98, 136–37, 165, 238, 246n, 247n, 249n, 250n, 255n, 257n
Haziness: and clarity in expressions, 91–92
Heidegger, Martin, 209
Heisenberg, Werner, 28, 48, 88, 99, 105, 161, 170, 191, 220, 235, 238, 256n; matrix mechanics, 212, 216; problems of natural language descriptions, 212–13; uncertainty principle, 85, 206. See also Bohr, Quantum theory
Heraclitus, 48, 52, 53, 85, 116
Hermeneutic loop, 229
Hernández, José, 195
Hertz, Heinrich, 84
Hesitation effect: in fantastic literature, 100
Hilbert, David, 64, 66, 67, 164, 250n. See also Paradox: Grand Hotel
Hinton, C. Howard, 81, 119, 134, 248n, 251n. See also Space-time manifold
Hobbes, Thomas, xiii
Hofstadter, Douglas, 72, 206, 262n; "authorship triangle," 33, 69; "strange loop," 169, 257n
Hologram, 66, 191, 250n, 258–59n; theory of mind and universe, 186. See also "The Aleph," Bohm, One-many, Unbroken wholeness
Holton, Gerald, 99
Hume, David, 2, 13, 14, 15, 26, 29, 135, 136, 138, 242
Hypercube, 119, 122. See also "Block" universe, Minkowski "block," Necker cube, Space-time manifold
Hyperfictionalization, 15, 30, 38, 44, 91–92, 95
Hypothetico-deductive method, xi–xii

"I": of "Borges and I," 92–93; disintegration of, 77
Ibarra, Nestor, xiii
I Ching, xiv, 174, 254n
Imaginary numbers, 99–100
Implicate-explicate, 183–88. See also Bohm, Enfolded-unfolded

Index

Incommensurability thesis, 210, 216–28. *See also* Radical meaning variance theory
Infinite regress: in "Circular Ruins," 34. *See also* Cantor sets, Consciousness, Paradox, Zeno
Infinitist: finitist versus, 56
Infinity, 22, 54; actual and potential, 48–49, 55–56; denumerable and nondenumerable, 55–56; and finitude, 109, 133; *horror infiniti*, 54; of nature, 192, 194; paradox of, 58; Pascal's fear of, 54; regarding quantum theory, 248–49n; Renaissance *coincidentia oppositorum*, 54; randomness regarding "The Lottery," 173–77; and totality of in "The Library," 71–72. *See also* "The Book," "The Library," "The Lottery," Cantor sets, Paradox, Zeno
Ingenieros, José, xiii
Intertextuality, 188–89, 193–94, 196, 207–8; not limited to textuality, 196–97
Ionesco, Eugene, 31
Irby, James E., 253n

James, William, 2, 237, 256n, 258n
Jeans, James, 97, 98, 108, 150–51, 166, 232, 258; universe as "great thought," 95, 158, 211. *See also* "Block" universe, Mathematics, Minkowski "block"
Johnson, Samuel, 158
Jordan, Pascual, 99
Jung, Carl G., xiii, 198

Kabbalah, xiv, 9, 81, 150, 171, 234, 258n
Kafka, Franz, 168, 189, 191, 237, 248n

Kant, Immanuel, 4, 8, 29, 49, 96, 125, 143, 161
Kasner, Edward, 61
Keats, John, 2–3, 8, 85
Kepler, Johannes, 214
Kierkegaard, Søren, 32, 189, 237
Kipling, Rudyard, xiii
Klein, Felix, 14
Kline, Morris, 14, 67, 68
K-mesons, 205. *See also* Quantum theory, Symmetry
Koch curve, 200–201
Koestler, Arthur, 88
Korzybski, Count, 31
Koyré, Alexandre, 86
Kuhn, Thomas S., 45, 216, 218, 228, 229, 247n, 262n. *See also* Incommensurability thesis, Paradigm, Radical meaning variance theory
Kummers, E. E., 247n

Labyrinth, 182, 196; and fourth dimension, 119–20; and infinite novel, 181–82. *See also* "Death," "The Garden," "The House," "The Library," "The Lottery," "The Secret"
Language problem, 213–28. *See also* Heisenberg
Laplace's superobserver, 123, 131–32, 252n. *See also* Newtonian-Cartesian machine model, Space-time manifold
Lasswitz, Kurd, 13, 128, 131–32, 247n. *See also* "The Library"
Layzer, David, 131–32, 254n, 263
Lehmann, Hugh, 18
Leibniz, Baron Gottfried Wilhelm von, 8, 11, 18, 207, 249n
Leucippus, 53
Light cone, 123–24, 146–47. *See also* Space-time manifold, Space-time singularity, Time-line, World-line

Linguistic idealism, 209
Literature of exhaustion, 243, 263n
Locke, John, xiii, 2
Logic of inconsistency, 18
Logocentrism, 250n, 255n
Lorich, Bruce, 38–39
Lucretius, 108, 109
Lugones, Leopoldo, 195, 261n
Lupasco, Stéphane, 31
Lyotard, Jean-François, 100, 253n

McCulloch, Warren, 211
Mach, Ernst, 28, 29, 161
Macintyre, Alasdair, 101
McTaggert, J. M. E., 137–38, 143
Magritte, René, 31, 71
Mallarmé, Stéphane, 235; dice throw, 171
Malraux, André, 189
Mandelbrot, Benoit, 200–201
Many-worlds interpretation: of quantum mechanics, 177–82, 192–93
Margenau, Henry, 102
Martínez Estrada, Enrique, 195, 261n
Mathematics: as convention, 19–21, 68; appropriateness for in "new physics," 89–90. *See also* Wigner
Matte Blanco, Ignacio, 64–65, 250n, 251n. *See also* Consciousness
Maurois, André, xi, 251n
Mauthner, Fritz, 161, 242, 263n; critique of language, 239–41
Maxwell, James Clerk, 84, 196–97
Maxwell's demon, 115, 130. *See also* Second law of thermodynamics
Meaning approximation, 224–25. *See also* Incommensurability thesis, Radical meaning variance theory

Meinong, Alexis, 1, 23–24, 30, 32, 165; mental objects, 24–25; in "Tlön," 27
Meyer, Leonard, 231
Mignolo, Walter, 180
Mill, J. S., 45
Miller, Arthur I., 105
Milleret, Jean de, 245n, 246n
Minkowski, Hermann, 121
Minkowski "block," 122, 123–24, 129, 141, 171, 180, 232, 251n, 252n, 261n. *See also* "The Library," Relativity theory, Time-line, World-line
Minotaur, 62–63
Mirrors. *See* Symmetry
Möbius strip, 125–27. *See also* Flatlander, Necker cube
Morgan, Augustos de, 14
Musil, Robert, 246n
Mystical insight: and "new physics," 149, 151–54

Nabokov, Vladimir, 68
Necessity-contingency, 133. *See also* Chance-necessity
Necker cube, 217, 256–57n; and perception, 126–27; and quantum superposition, 158. *See also* Möbius strip
Neruda, Pablo, 106
Newman, James, 61
Newton, Isaac, 131, 183, 198, 214, 241; Newtonian worldview, 114
Newtonian-Cartesian machine model, 42, 98, 113–14
Newtonian physics: versus Einsteinian physics, 218–19, 225, 238; "simultaneity," 218, 225; "mass," 219. *See also* Incommensurability thesis, Radical meaning variance theory
Nietzsche, Friedrich, 18, 31, 84, 105, 113, 116, 118, 134, 171, 195, 198, 209, 238, 242

Index

Nominalism, 1–4, 157–58; regarding computer thought, 11–12; realist-nominalist controversy, 1–5
Northrop, F. S. C., 29

Occam, William of, 11
One-many, 133, 151, 152, 154, 176, 191. *See also* "The Aleph," Hologram, "New Refutation," "The Zahir"
Otto, Rudolf, 152–53
Ouspensky, P. D., 142, 230

Paracelsus, 262n
Paradigm, 219–20. *See also* Conceptual framework, Incommensurability thesis
Paradox, 32–52; the age of, 31; in "Avatars," 42, 43, 45, 49; in "The Book," 56–57; Borges's penchant for, xiii; in "Circular Ruins," 33–38; Copernicus's, 214; in "Death," 46–48, 203–5; defined, 32, 39; Einstein-Podolsky-Rosen (EPR) thought experiment, 168–69; in "The Garden," 177–82; regarding Gödel's proof, 67; Hilbert's Grand Hotel, 64, 66; holographic, 186; of infinity, 58–60; liar, 39, 67, 68, 69, 71; lottery, 174–75; in "The Lottery," 118–19, 173–77; map, 186; *Meno*, 191; prisoner, 40; in quantum theory, 187; Russell's, 165–66, 169, 191; in "Tlön," 159–60, 164–66; Zeno's, 42–52
Park, David, 115
Parmenides, 42, 48, 53, 109, 116, 117, 143
Pascal, Blaise, xiii, 54, 112
Pauli, Wolfgang, 161
Peano, Giuseppe, 14
Peirce, Charles S., 14, 172–73
Penrose, Roger, 261n

Physics: of chaos, 246n. *See also* Dissipative structures
Piaget, Jean, 4
Picasso, Pablo, 163
Pirandello, Luigi, 31, 68, 163
Planck, Max, 105, 162, 215
Plato, 10, 105, 109, 152, 191; Plato-Aristotle distinction, 246n; Platonic forms, 223
Poincaré, Henri, 88, 109, 250n
Polanyi, Michael, 45, 216, 225, 228, 241n, 249n
Pollock, Jackson, 215
Popper, Karl, 216–17, 218, 262n; hypothetico-deductive method, 88; myth of the framework, 262n
Postmodernism, 252n
Pribram, Karl, 259n. *See also* Hologram
Prigogine, Ilya, 254n
Ptolemy, 101
Pynchon, Thomas, 246n
Pythagoras, 85, 113; Pythagoreans, 95, 105

Quantum theory: ambiguity and uncertainty, 162–65; asymmetry, 205–6; no consensus on, 156; continuous and discretized space, 187; probability, 156–57; problem of natural language description, 210; superposition, 160. *See also* Aleph-Zahir complementarity, Bohm, Bohr, Copenhagen interpretation, Heisenberg, Schrödinger, Wheeler, Wigner
Quaternions, 98–99
Quixote, 70, 129, 166, 172, 192, 194, 195, 207

Radical meaning variance theory, 216–28, 236–37, 238, 239, 261–62n. *See also* Incommensurability thesis, Paradigm

Index

Random: music, 215; books in the "Library," 128–29
Real-irreal, 106–8, 118–19, 189; in quantum theory, 161
Realist-nominalist controversy, 1–5, 209
Reimann, Georg F. B., 14
Reimannian geometry, 23
Relativity theory, chaps. 4, 5. *See also* Aleph-Zahir complementarity, Black holes, "Block" universe, "The Aleph," "Death," Einstein, Minkowski "block," Space-time manifold, Space-time singularity, Time-line, World-line
Rescher, Nicholas, 18, 21, 25, 31, 247n
Rest, Jaime, 3
Robbe-Grillet, Alain, 68
Rodríguez, Monegal, Emir, 107, 202, 239
Ross, Waldo, 49
Royce, Josiah, 251n
Rucker, Rudy, 127, 145, 153, 170, 253n, 254n
Russell, Bertrand, 3, 11–12, 14, 24, 28, 31, 44, 166, 191, 250n

Sábato, Ernesto, 203
Sarraute, Natalie, 68
Schnitzler, Arthur, 252n
Schopenhauer, Arthur, 2, 66, 78, 81, 106, 116, 119, 138, 244, 255n
Schrödinger, Erwin, 78, 80, 105, 150, 153–54, 158, 206; on consciousness, 167; wave mechanics, 212, 216
Schrödinger's cat, 157–58, 160–61, 177, 179, 181
Second law of thermodynamics, 115, 170. *See also* Eternal return, *Historia de la eternidad*
Semiology: Saussurean, 194
Serres, Michel, xiii–xiv

Set theory, 60–61. *See also* Cantor sets, Infinity
Shakespeare, William, 78, 101, 189
Shelley, Percy B., 85, 190
Space-time manifold, 111–12, 119–24, 232
Space-time singularity, 145–46
Spengler, Oswald, 19, 130, 244
Spinoza, Baruch, 2, 8, 203, 249n
Stevenson, Robert Louis, xiii
Sturrock, John, 3
Sucre, Guillermo, 32
Sufi thought, xiv
Sunyata, 58
Superstrings, 246n
Symmetry, 198–208; rotational, mirror, bilateral, 198–200; and time, 198. *See also* "Circular Ruins," "Death," K-mesons, Quantum theory

Tacit knowing, 225
Tagore, Rabindranath, 107
Tarski, Alfred, 31
Taylor, Charles, 229, 262n
Theory of logical types, 63. *See also* Paradox: Russell's
Thought experiments: "as if" hypostats, 15, 16–17. *See also* Vaihinger
Time-line, 143, 146, 171. *See also* World-line
Todorov, Tzvetan, 100
Toulmin, Stephen, 19
Tragedy-comedy: in "Averroes' Search," 223

Unamuno, Miguel de, 67
Unbroken wholeness, 183–88, 259–60n. *See also* Bohm, Quantum theory, Relativity theory
Upanishads, 66, 255n

Index

Vaihinger, Hans, 1, 3, 22, 23, 25–28, 30, 32, 45, 48, 238, 246n; "as if" hypostats, 15, 16–19
Valéry, Paul, 190, 207
Verne, Jules, 107
Vosburg, Nancy, 188

Wahl, Jean, 2
Waismann, Friedrich, 214
Wells, H. G., xiii, 107, 145
Wheeler, John Archibald, 88, 169, 170, 172, 177, 180, 181, 189, 192; participatory universe, 167–68. *See also* Quantum theory
Wheelock, Carter, 3, 5, 16, 255n
Whitehead, Alfred North, 14, 34, 48, 98, 105, 139–40, 141, 184
Whitman, Walt, 162
Whitrow, G. J., 249n
Whittaker, Edmund, 99
Wigner, Eugene, 90, 205; consciousness in quantum reality, 167; quantum solipsism, 181, 257n
Wilkins, John, xiii, 22, 207
Winch, Peter, 229, 262n
Wittgenstein, Ludwig, x, 3, 19, 20–21, 68, 105, 119, 220, 230–31, 247n, 262n
Word magic, 7, 14
World-image, 233
World-line, 122–23. *See also* Minkowski "block," Space-time manifold, Time-line

Xenophanes, 108

Zen Koans, 163
Zeno, xiii, xvi, 31, 32, 56, 58–59, 80, 85, 106, 109, 112, 137, 139, 147, 159–60, 163, 182, 187, 189, 237, 248n; Cantor sets, 60; paradoxes, 42–52; related to quantum theory, 249n. *See also* Infinity, Paradox
Zero: concept of, 58